Lecture Notes in Energy

Volume 51

Lecture Notes in Energy (LNE) is a series that reports on new developments in the study of energy: from science and engineering to the analysis of energy policy. The series' scope includes but is not limited to, renewable and green energy, nuclear, fossil fuels and carbon capture, energy systems, energy storage and harvesting, batteries and fuel cells, power systems, energy efficiency, energy in buildings, energy policy, as well as energy-related topics in economics, management and transportation. Books published in LNE are original and timely and bridge between advanced textbooks and the forefront of research. Readers of LNE include postgraduate students and non-specialist researchers wishing to gain an accessible introduction to a field of research as well as professionals and researchers with a need for an up-to-date reference book on a well-defined topic. The series publishes single- and multi-authored volumes as well as advanced textbooks.

Indexed in Scopus and EI Compendex The Springer Energy board welcomes your book proposal. Please get in touch with the series via Anthony Doyle, Executive Editor, Springer (anthony.doyle@springer.com)

More information about this series at https://link.springer.com/bookseries/8874

Clemens van Dinther · Christoph M. Flath ·
Reinhard Madlener
Editors

Smart Grid Economics and Management

Springer

Editors
Clemens van Dinther
Information Systems and Business Analytics
ESB Business School at Reutlingen University
Reutlingen, Germany

Christoph M. Flath
Chair of Information Systems and Business Analytics
Julius-Maximilians-Universität Würzburg
Würzburg, Germany

Reinhard Madlener
Chair of Energy Economics and Management
School of Business and Economics / E.ON Energy Research Center
Institute for Future Energy Consumer Needs and Behavior (FCN)
RWTH Aachen University
Aachen, Germany

ISSN 2195-1284 ISSN 2195-1292 (electronic)
Lecture Notes in Energy
ISBN 978-3-030-84284-0 ISBN 978-3-030-84286-4 (eBook)
https://doi.org/10.1007/978-3-030-84286-4

This Springer imprint is published by the registered company Springer Nature Switzerland AG
The registered company address is: Gewerbestrasse 11, 6330 Cham, Switzerland

Foreword

Digital Energy Systems are characterized by their high degree of complexity in monitoring and controlling a vast number of energy consumers and producers over a dynamically optimized infrastructure in synchronization and accordance with social behavior orchestrated and governed through internationally run markets. From a technical perspective, digitalized energy systems may be described as a Cyber-Physical System (CPS) that has originated from the interconnection of electromechanical components with IT-interfaces and IT-functionalities, closely linked with the surrounding physical world and its ongoing processes, providing and using, at the same time, data-accessing and data-processing services available on the Internet. In other words, CPS can be generally characterized as physical and engineered systems whose operations are monitored, controlled, coordinated, and integrated by a computing and communicating core. The term *cyber-physical system* has been created in response to the need for a new theoretical basis for the research and development of large, distributed, complex systems. In that vein, cyber-physical *energy* systems exhibit special characteristics that uniquely set them apart from other CPS:

- They are regarded as a critical infrastructure (CRITIS) and indispensable lifelines of modern societies.
- They have cross-continental size, e.g., in Europe from North Africa to Scandinavia, from Ireland to Asia.
- They show instantaneous propagation velocity of dynamic phenomena, e.g., imbalances or instabilities.
- Their actors exhibit omnipresent conflicts of objectives, e.g., monetary, technical, environmental, and (national/international) political interests.
- They are undergoing rapid and fundamental change due to the energy transition and digitalization.
- Energy systems include a distinct market perspective, i.e., system-wide orchestration and resource provisioning through economic incentives and specialized market models.

Following this reasoning, smart grid research, development, and engineering requires and specializes the need for a theoretical basis for interdisciplinary education, research, and development of digitalized energy systems—significantly extending the scope of CPS.

Competences necessary for mastering this novel field are not only in the general area of *system intelligence*—algorithms for adaptive control and continuous dynamic optimization of the complex and expansive international power supply system—and the knowledge of methods to create and orchestrate *overall system competence*—complexity control through decomposition and abstraction, identification of and focus on generalizable principles, detection of decoupling points for effective governance, and avoidance of bottlenecks. Smart grid education also requires fundamentals in market principles and design of business cases following regulatory and institutional system aspects as well as reflecting the special requirements of trading electric power.

Clemens, Christoph, and Reinhard have authored and edited a text book that bridges a significant part of the gap within the educational literature in ICT-reliant smart grids—the scope ranges from smart grid economics, market models and interaction mechanisms to regulatory aspects, and modeling and data analytics in smart grid systems. Complemented with exercises as well as business and use cases adopted from real-world examples, this textbook will become a valuable companion for students, researchers, and practitioners in this future-oriented topic. It will certainly find room in many places in my lectures.

Oldenburg, Germany
February 2022

Sebastian Lehnhoff

Preface

The idea for this textbook on Smart Grid Economics and Management goes back to a joint workshop held in Karlsruhe in 2011, where Hans-Jürgen Appelrath (†) proposed to jointly write a textbook. Since then, the concept realization and many details of the book have evolved continuously, and eventually a relatively large number of authors have contributed to different chapters of the book.

All three editors have been teaching on the subject area for years, in part even jointly, at their home universities. This enabled us to put together, and regularly update, the teaching material. At the same time, we realized that such a textbook did not exist yet but was in urgent need—somewhere between energy economics textbooks (such as Zweifel et al., 2017) and power economics (such as Kirschen and Strbac, 2004, or Stoft, 2002), but tackling in more detail the roles of Information and Communication Technologies (ICT), smart grids, distributed resources, and demand response.

The book covers altogether eight subject areas. Chapter "Energy Systems Today and Tomorrow" (by Clemens van Dinther and Reinhard Madlener) is the introduction to the entire book and field covered. It provides a broader overview on the energy transition, driven both by climate change mitigation (decarbonization) and technical change (digitalization), and leading to more decentralization of the energy supply systems as well as democratization (the so-called "4 Ds").

Chapter "Smart Grid Economics" (by Reinhard Madlener) introduces Smart Grid Economics, addressing the issues at hand when dealing with smart grids from an economics perspective that are beyond standard energy economics textbooks. By doing so, it paves the ground for several other chapters.

Chapter "Demand Side Management" (by Joan Batalla-Bejerano, Elisa Trujillo-Baute, and Reinhard Madlener) gives an overview of demand side management in the smart grid context, dealing with potentials, consumer participation, flexibility markets and new business opportunities (e.g., by aggregating demand response potentials), and dynamic pricing for final consumers.

Chapter "Market Engineering for the Smart Grid" (by Philipp Staudt and Christof Weinhardt) tackles the field of energy market engineering as a scientific method. They show that energy systems need to be tackled from a temporal and spatial perspective, what new challenges are created by high shares of intermittent renewable power generation, and how smart grids can help to alleviate some of the problems there are. An overview is given on how markets can be designed to leverage the benefits that can be created by smart grids.

Chapter "Regulatory and Institutional Aspects of Smart Grids" (by Gert Brunekreeft, Marius Buchmann, and Anna Pechan) covers the regulatory and institutional aspects relevant to smart grids. Here, governance models, network regulation, and grid use pricing are discussed in detail, and the changes compared to conventional electricity market governance, regulation, and pricing models are made clear. The latter is typically based on capacity scarcity, in the smart grid context in a highly disaggregated manner (households or even distributed smart devices) rather than on grid areas (as in the case of classical nodal or zonal pricing).

Chapter "Modeling Smart Grid Systems" (by Dominik Möst et al.) is the chapter for modelers who want to learn and better understand how to model smart grid systems. The chapter gives an useful overview of what the goals and dimensions of modeling are, and many of the specifics that need to be known and properly understood to be able to build and make good use of models dedicated to particular needs ("horses for courses").

Chapter "Smart Grid Analytics" (by Christoph M. Flath and Nikolai Stein) addresses predictive and prescriptive analytics techniques for smart grid systems. It is an introduction to machine learning applied to smart grid topics and also provides several application scenarios.

Chapter "Business Model Design" (by Markus Lau, Clemens van Dinther, and Orestis Terzidis) deals with business model development and business and use cases in the context of business enabled by digitalization (i.e., Information and Communication Technologies, ICT) in general, and smart grids in particular.

Chapter "Case Studies in the Smart Grid Sector" (by Clemens van Dinther, Markus Lau, and Orestis Terzidis) takes up examples of two business models that have emerged in the context of the smart grid. These are exemplary for a multitude of new business ideas and illustrate the spectrum of economic transition.

An important element of the textbook are exercise questions and some business and use cases adopted from the real world.

The book is primarily aimed at students and academics in the fields of economics, management science, information science, electrical engineering as well as various kinds of consultants and other practitioners.

Needless to say that the first edition of such a textbook offers tremendous scope for improvement, especially on such a still novel and very dynamically evolving field. Therefore, we do hope that many readers will write us their comments, pointing out shortcomings and new ideas (e.g., for additional literature and topics to be covered).

We are grateful for the contributions of many colleagues in terms of (parts of) individual chapters without which the publication of this book would have been delayed further. Our thanks go to Hans-Jürgen Appelrath (†) who provided the impetus for this book, to Sebastian Lehnhoff, who accompanied the book at the beginning and was available for many discussions, and to the many teaching assistants and students attending our courses and providing valuable inputs. We also want to thank our families for their patience, and our Managing Editor from Springer and the whole team for fruitful suggestions and efficient interaction.

Reutlingen, Germany Clemens van Dinther
Würzburg, Germany Christoph M. Flath
Aachen, Germany Reinhard Madlener
February 2022

References

Kirschen, D. and G. Strbac (2004). Fundamentals of Power Systems Economics, Wiley.
Stoft, S. (2002). Power System Economics: Designing Markets for Electricity. Wiley-IEEE Press.
Zweifel, P., A. Praktiknjo and G. Erdmann (2017). Energy Economics - Theory and Applications, Springer Texts in Business and Economics, Springer, Berlin, Heidelberg.

Contents

Contributors

Joan Batalla-Bejerano Funseam and Chair of Energy Sustainability IEB-UB, Barcelona, Spain

Gert Brunekreeft Bremen Energy Research, Jacobs University, Bremen, Germany

Marius Buchmann Bremen Energy Research, Jacobs University, Bremen, Germany

Clemens van Dinther ESB Business School, Reutlingen University, Reutlingen and VIVAVIS AG, Ettlingen, Germany

Christoph M. Flath Chair of Information Systems and Business Analytics, Julius-Maximilians-Universität Würzburg, Würzburg, Germany

Hannes Hobbie Chair of Energy Economics, Technische Universität Dresden, Dresden, Germany

Markus Lau Institute for Entrepreneurship, Technology Management and Innovation (EnTechnon), Karlsruhe Institute of Technology (KIT), Karlsruhe, Germany

Reinhard Madlener Chair of Energy Economics and Management, School of Business and Economics / E.ON Energy Research Center, Institute for Future Energy Consumer Needs and Behavior (FCN), RWTH Aachen University, Aachen, Germany

Steffi Misconel Chair of Energy Economics, Technische Universität Dresden, Dresden, Germany

Dominik Möst Chair of Energy Economics, Technische Universität Dresden, Dresden, Germany

Anna Pechan Bremen Energy Research, Jacobs University, Bremen, Germany

David Schönheit Chair of Energy Economics, Technische Universität Dresden, Dresden, Germany

Philipp Staudt Institute of Information Systems and Marketing, Karlsruhe Institute of Technology (KIT), Karlsruhe, Germany

Nikolai Stein Chair of Information Systems and Business Analytics, Julius-Maximilians-Universität Würzburg, Würzburg, Germany

Orestis Terzidis Institute for Entrepreneurship, Technology Management and Innovation (EnTechnon), Karlsruhe Institute of Technology (KIT), Karlsruhe, Germany

Elisa Trujillo-Baute University of Lleida and Chair of Energy Sustainability IEB-UB, Barcelona, Spain

Christof Weinhardt Institute of Information Systems and Marketing, Karlsruhe Institute of Technology (KIT), Karlsruhe, Germany

Christoph Zöphel Chair of Energy Economics, Technische Universität Dresden, Dresden, Germany

Acronyms

AI	Artificial intelligence
ANN	Artificial neural network
BEV	Battery electric vehicle
BM	Business model
BRP	Balance responsible party
CEC	Citizen Energy Communities
CEER	Council of European Energy Regulators
CHP	Combined heat and power
CMS	Central market system
CO_2	Carbon dioxide
CPP	Critical peak pricing
DA	Day-ahead
DER	Distributed energy resources
DG	Distributed generation
DR	Demand response
DSM	Demand side management
DSO	Distribution system operator
EPC	Energy performance contracting
ERP	Enterprise resource planning
ESC	Energy supply contracting
ESCO	Energy service company
ESI	Energy system integration
ESM	Energy supply model
ETS	Emissions trading scheme
EV	Electric vehicle
GA	Genetic algorithm
GHG	Greenhouse gases
GIS	Geographic Information System
HVAC	Heating, ventilation, and air conditioning
ICT	Information and communication technology
IDO	Independent distribution operator

IDSO	Independent distribution system operators
IL	Interruptible load
IoT	Internet of things
IPCC	Intergovernmental Panel on Climate Change
IRENA	International Renewable Energy Agency
ISO	Independent system operator
IT	Information technology
ITO	Independent transmission operator
LCoE	Levelized cost of energy
MAD	Mean absolute deviation
ML	Machine learning
MSE	Mean squared error
Mtoe	Million tons of oil equivalent
NILM	Non-intrusive load monitoring
NUTS	nomenclature des unit´es territoriales statistiques
P2P	Peer-to-peer
PBR	Performance-based regulation
PCA	Principle component analysis
PHEV	Plug-in hybrid electric vehicle
PIM	Performance incentive mechanism
PtX	Power-to-X
PV	Photovoltaics
RES	Renewable energy sources
RTP	Real-time pricing
SG	Smart grid
SNG	Synthetic natural gas
SVM	Support vector machine
SVR	Support vector regression
TCO	Total cost of ownership
ToU	Time of use
TSO	Transmission system operator
VER	Variable energy resources
VPP	Virtual power plant

List of Figures

Energy Systems Today and Tomorrow

Smart Grid Economics

Demand Side Management

Market Engineering for the Smart Grid

Regulatory and Institutional Aspects of Smart Grids

Modeling Smart Grid Systems

Smart Grid Analytics

Business Model Design

List of Tables

Energy Systems Today and Tomorrow

Clemens van Dinther and Reinhard Madlener

Learning Objectives

- Be able to understand the importance and types of energy systems,
- Be able to grasp the drivers, barriers, and challenges of the ongoing sustainable energy systems transition,
- Be able to classify and identify dimensions of smart grids (special role of power systems and ICT),
- Be able to describe the relevance of technological diffusion (social welfare perspective, path dependence/lock-in, complexity, supply-push/demand-pull),
- Be able to explain key policy aspects related to energy systems transition (policy instruments, regulation, market failure, uncertainty)

The energy systems are under tremendous change, driven both by technical change (digitalization) and climate protection policy. Overall, it is expected that the future energy system will be decarbonized, digitalized, distributed—and thus also democratized (the four "Ds"). An increased digitalization and the vast system transformation also bear new risks, some of which are related to cybersecurity.

C. van Dinther (✉)
ESB Business School, Reutlingen University, Alteburgstr. 150, 72762 Reutlingen, Germany
e-mail: clemens.van_dinther@reutlingen-university.de

R. Madlener
Chair of Energy Economics and Management, School of Business and Economics / E.ON Energy Research Center, Institute for Future Energy Consumer Needs and Behavior (FCN), RWTH Aachen University, Aachen, Germany
e-mail: RMadlener@eonerc.rwth-aachen.de

C. Dinther et al. (eds.), *Smart Grid Economics and Management*,
Lecture Notes in Energy 51, https://doi.org/10.1007/978-3-030-84286-4_1

1 Energy Systems—An Essential Element of Society

Energy is a key factor for human development and economic growth (Stern 2011). It appears in different forms (e.g., potential, chemical, electric, kinetic, or thermal energy). Apart from its physical characteristics, energy has also an economic, societal, and environmental dimension. No matter which area we look at in our everyday lifes, energy plays an important role.

The downside of massive (fossil) energy consumption is the increasing emission of carbon dioxid (CO_2) or greenhouse gases (GHG), respectively, as well as environmental pollution. The Intergovernmental Panel on Climate Change (IPCC) published a First Assessment Report in 1990 IPCC (1990, p. xi) stating that "there is a natural greenhouse effect which already keeps the Earth warmer than it would otherwise be" and "emissions resulting from human activities are substantially increasing the atmospheric concentrations of greenhouse gases [...]" enhancing "[...]the greenhouse effect, resulting on average in an additional warming of the Earth's surface." Twenty-five years later, 195 nations have fixed an agreement on the mitigation of GHG emissions. This agreement is known as the Paris Agreement (European Union 2016) and seeks to limit the global mean temperature increase to 2 °C and mitigate the increase at 1.5 °C. In order to reach this objective, each country presented comprehensive climate safety plans. As a result, most Western industrial countries have pledged to radically transform all energy-consuming sectors. This transformation process that goes well beyond changing the technical systems is known as sustainable energy transition. This has major effects also on the electricity system where conventional electricity generation from fossil fuels needs to be replaced by generation from renewable resources.

The need for a sustainable energy transition has implications for many areas of the energy industry. In this book, we focus on the ramifications for the electricity sector by looking at the future of distribution grids, the so-called *smart grid (SG)*. In the context of this transition, it is often referred to the Ds of the energy transition. There is no clear original scientific source for these Ds. However, there typically is an agreement on the first three Ds (e.g., Agora 2017)—decarbonization, digitalization, and decentralization. Other Ds include democratization, the dominance of fixed costs, or degression of technology costs (cf. also Chapter 6).

The *decarbonization* of electricity generation leads to a more *decentralized* structure of the electricity system, as the renewable energy generation plants, i.e., wind turbines and PV, are smaller and are installed across many locations. Generation-led system control is insufficient in this setting and in turn we need more flexibility on the demand side. This flexibility will not be achieved with manual processes or behavioral changes. New technology is necessary to control devices, to meter actual consumption or monitor the grid. This calls for *digitalization* to make the distribution grid smart. Controllable devices, consumer-side generation, and the need for demand-side flexibility empower consumers and turn them into prosumers. In turn, we see a *democratization* of the electricity system. The latter is not only a technical challenge. In fact, while the structure of the system is changing and new players are emerging, regulators will need to adapt market mechanisms to reflect changing cost structures. Similarly, network management is changing, new regulation is required and new business models are emerging. Renewable power generation is characterized by the *dominance of fixed cost*, although we observe a *decrease of the technology cost* due to technological progress and scale effects.

In this new textbook, we cover these arising challenges and mitigation strategies. We thereby focus on economics and management, new trends, information processing, and entrepreneurship as well as on the modeling of "smart" energy systems, rather than on technical aspects.

Electrical energy is generated from different sources. These include primary energy sources such as oil, coal, gas, nuclear material, water, or renewable sources. Worldwide primary energy consumption increased rapidly in the last years, up to 14,126 million tons of oil equivalent (Mtoe) in 2017 (Worldbank 2018). Oil has the largest share with 34.2%, followed by coal with 27.6% and natural gas at 23.4%. Nuclear (4.4%), hydro-electric (6.8%) and renewable sources (3.6%) have smaller shares. In 2017, worldwide electricity production was 25,551.3 TWh (or 2,197 Mtoe). Therefore, presently about 15% of the world's energy consumption is electrical.

The knowledge of humans to physically transform (fossil) energy at large scale - in order to use it for industrial and manufacturing processes started the industrial revolution in the eighteenth and nineteenth centuries. This led to economic growth and an unprecedented societal development change of life. Today, Western economies as well as the developing and emerging economies cannot work without reliable energy supply. Although there is a relationship between energy consumption and economic growth, a synchronous development can no longer be observed since the 1970s. Instead, we have seen a decoupling of GDP and energy consumption (Banschbach 2003).

As already mentioned, energy can exist in different forms. Primary energy sources are crude oil, natural gas, coal (lignite and hard coal), nuclear raw materials, and renewable energy sources such as hydropower, solar energy, wind energy, geothermal energy, or biogas. The primary energy is transformed depending on the application, e.g., into electrical energy, heat (heating, process heat), or for the transport sector. The proportion of primary energy sources used varies in different regions of the world. According to BP's energy report, oil is the most frequently used primary energy, with a share of approximately one third of global consumption, followed by coal with a share of 26%. Asia is the largest consumer of coal and oil, while North America leads the rankings for nuclear energy and natural gas. Despite a steady increase, renewable energies still have the smallest share worldwide at 3.6%. In South and Central America, hydropower plays an important role. In Europe, the share of renewable energies is the largest across all industrialized nations.

In Germany, approximately 36% and in the USA 38% of primary energy is transformed into electrical energy. In Germany, electrical energy is the most important form for industry (31%), commerce/trade/services (36%) and households (19%) alongside natural gas. When we referring to the energy industry and the energy system in this textbook, typically the electricity system is meant.

The energy industry is faces with the challenge of providing electricity economically, ecologically, and reliably. We are therefore also talking about the energy policy trilemma of security of supply (social dimension), affordability (economic dimension), and environmental impact (ecological dimension). Since electricity, as an essential commodity for the economy and society, is vital for survival, all three dimensions are important. However, the question arises as to the political priorities, since the goals are interdependent. For example, redundancies in generation and transmission are necessary to reduce the risk of system failure. A high level of supply security leads to higher costs and thus has a negative impact on profitability. The ecological dimension also influences economic efficiency.

Electricity is a special economic commodity in several respects.

1. It is a *homogeneous good*, i.e., electricity cannot be distinguished physically. In principle, quality criteria can be defined, e.g., with regard to reliability or also with regard to generation source, frequency, voltage, and current, but electrons are electrons. In contrast to the homogeneous good "electricity," the electricity system is extremely heterogeneous. There are differences in the generation technology, in the structure of the system or with regard to the economic framework conditions which are laid down by law. Due to the heterogeneous system architecture/components, additional services and market mechanisms are necessary.
2. Power generation and consumption must be *continuously balanced*. For this reason, the electricity system was designed so that large power plants were built close to the consumption centres.
3. Electricity transport and distribution are *natural monopolies* due to the combination of high fixed investment costs and negligible marginal transport costs.[1]
4. Electricity *cannot be stored* easily or without losses.[2] In addition, electricity cannot be routed. Governed by physical laws, the current will flow along the path of least electrical resistance.

A working power system is indispensable. The structure of electricity systems differs worldwide, not only in terms of their technical structure but also in terms of their economic role and organization. Various factors must be taken into account in the structure of the electricity system. Geography, regulation, economy, and technology play a role in the structure of the electricity system. For example, a carbon-neutral energy system has more decentralized generation plants and other distributed resources, and thus also requires more intelligent control due to decentralisation.

From the perspective of the electricity grid, the requirements typically differ for differing population densities: Rural regions typically have a low population density. This means that long lines are necessary to connect buildings. Depending on the terrain, the grid connection can be complex if, for example, mountains or water bodies have to be overcome. In urban areas, the settlement density is greater. As

[1] In economics, one speaks of natural monopolies when sub-additive costs exist, i.e., the costs of a monopolist are smaller than the sum of the costs of all competitors.

[2] Storing electrical energy is only possible by transforming it into another form of energy, such as kinetic, chemical, or gravitational energy, and is therefore very expensive.

the number of buildings and appliances increases per square meter, so does energy demand, which must be covered by the capacity of the grid. Of course, the level of economic development and the type of buildings is also important—industrial or commercial areas have greater energy demand than residential areas. This results in heterogeneous requirements for the electricity grid, especially against the background of growing decentralized generation capacity. For example, isolated solutions are becoming attractive for sparsely populated regions, while in densely populated regions, intelligent coordination of generation and consumption may be a more viable option.

If we look at the geographical situation in Europe and compare it with other continents, we can see clear differences. A good indicator is population density. The lower the population density, the less demand there is, i.e., with increasing decentralization in generation, intelligent grids, i.e., the possibility of grid control measures, are more appropriate in more densely populated areas. Due to its importance for the national economy, the energy sector has always enjoyed special attention. In the twentieth century, both in the USA and in Europe, the energy supply was secured by governmental organizations. In the 1980s, the discussion about the liberalization of the sector began and, in the 1990s, it was pushed forward in different forms on both continents. Due to high investment costs and low marginal costs, energy networks are natural monopolies that remain under state regulation even after market liberalization. Generation, however, has been privatized and is mainly competitive. However, the approaches in the USA and EU differ. While the USA focuses on locational marginal pricing, the EU has implemented a stock exchange-oriented market. Both models have advantages and disadvantages which we do not want to discuss here any further, however, but instead want to refer the reader to the standard energy economics textbooks.

2 Drivers, Barriers, and Challenges of the Ongoing Energy Systems Transition

The future of energy supply is green, i.e., electricity is more and more produced from renewable production technologies such as PV, wind, water, and biogas (synthetic natural gas (SNG), respectively) instead of using fossil generation plants. This trend has several reasons, one is the rapid decline in renewable energy costs and another is the aim to reduce carbon dioxide emissions. In the Paris Agreement, the international community agreed to reduce GHG (especially CO_2) emissions in order to slow down climate change. Many countries are discussing or taking measures to decarbonize their economies, i.e., replace fossil fuels with renewable energies. This means not only that conventional power plants (coal-fired power plants, nuclear power plants) must be substituted for electricity production but also that other sectors such as transport or industry must be converted from fossil fuels like oil and gas to electricity. As a result, the overall demand for electricity is rising, which makes a massive

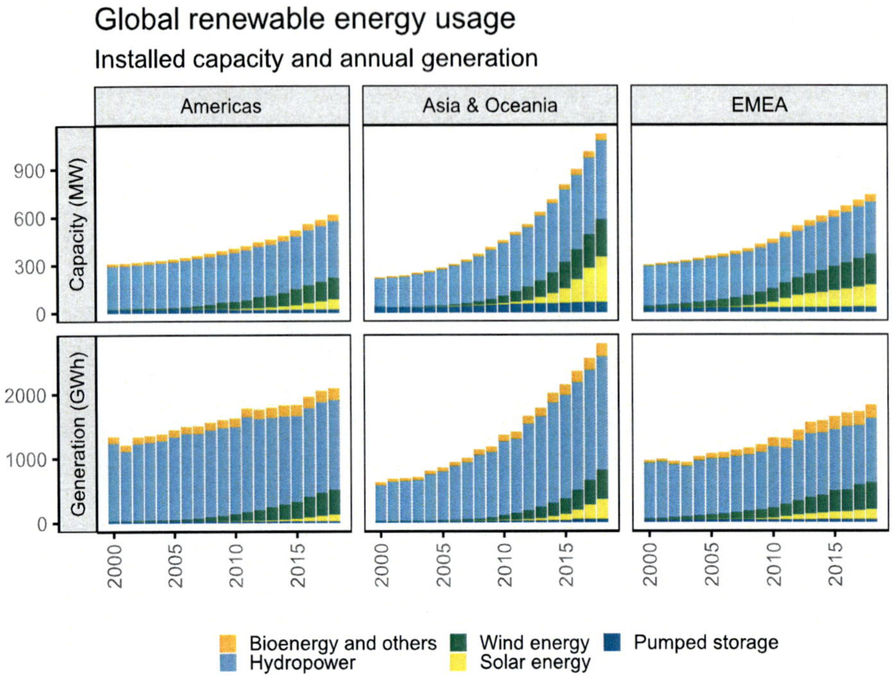

Fig. 1 Global installed renewable generation capacity

expansion of renewable power generation plants and transmission grids necessary. The worldwide installed capacity of PV has grown enormously in the last 20 years. In the period from 2008 to 2018, we see a tenfold increase in worldwide installed capacity. Figure 1 shows the development from 2000 to 2017. We are observing strong growth for Asia (especially China), Europe, and North America. PV yield is dependent on solar irradiation. Due to the geographical location, the largest yields are in the south of North America, Central and South America, Africa, and Oceania. With regard to these areas with high solar irradiation, it is interesting to note that the growth of solar PV in these regions has been relatively low over the period considered. However, the use of solar energy in these regions would be reasonable and is also being discussed, for example, for the production of regeneratively produced hydrogen.

Given the technological advances which also result in cost reduction a growth of installed capacity is expected. According to IRENA (2020), a study undertaken by the International Renewable Energy Agency (IRENA), the PV capacity could rise up to 8,500 GW by 2050 (Fig. 3).

Technological advances have led to considerable cost reductions in solar PV-electricity. Figure 2 shows average total installed costs, capacity factors, and LCoE for the period from 2012 to 2018. In 2010, the cost were still at 37 Cts/kWh, whereas in 2018, the kWh is produced at 9 Cts. This is a decrease by 75%. This could have been achieved through an increase in efficiency. Dependent on the location of the

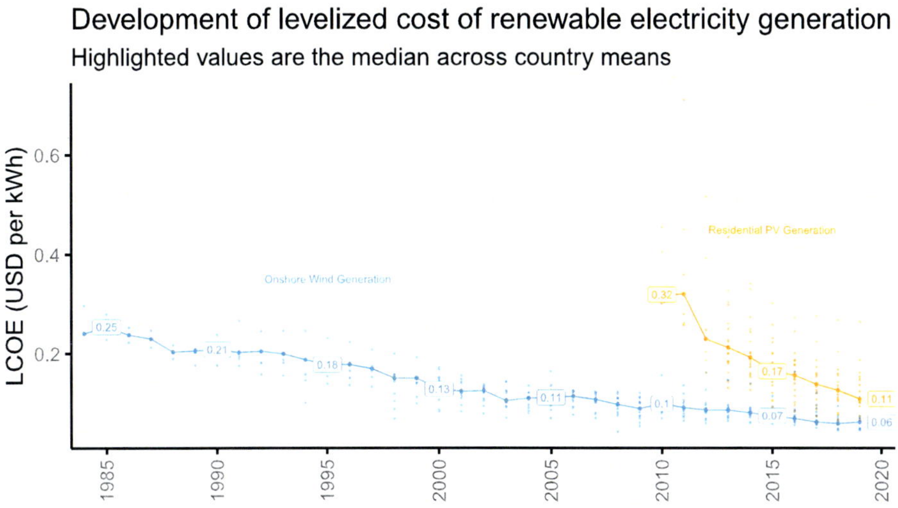

Fig. 2 Global weighted average LCoE for onshore wind and residential solar PV

Solar PV generation plant the LCoE could be lower at approx. 5 Cts/kWh, which makes it competitive with other conventional power generation technologies.

What we observe for solar PV is also true for wind turbines which have developed well in the past decades with respect to installed capacity, cost, and efficiency. Worldwide installed capacity has increased from 16 GW in 2000 up to 622 GW in 2019. As for solar PV, we see the largest increase in Asia (mainly China), Europe, and North America. Figure 3 shows the development. The large increase in installed capacity has also to do with the technological advances. Walker (2020, p. 410) analyzes the trend in turbine capacity. He remarks that the average turbine in the USA in 1990 was 0.218 MW; in 2005, it has risen up to 1.5 MW and increased further to 2.56 MW in 2019. In 2016, the largest turbine has a capacity of 6 MW. Wind blows more reliably with increasing distance from the ground. This is why the towers for wind turbines are getting higher and higher. General Electric, for example, is planning a 12-MW turbine with 260 m hub height above the earth's surface as an off-shore plant. With a rotor blade of 107 m length the turbine even towers above the Empire State Building in New York (which is 319 m of height). The first turbine of this type was produced in 2019 in France.

According to the efficiency increases of wind turbines, we also observe a significant drop in turbine prices and project cost which result in overall declining LCoE. Figure 2 gives an overview on the weighted average cost development during the 2010s. According to IRENA (2019, p. 19), the cost span of the 5–95% percentile in 2010 was between 5.6 and 11.4 Ct/kWh and declined to 4.3–9.9 Ct/kWh in 2018. Costs differ by region, and are dependent on production and implementation facilities. Costs for off-shore plants are still higher and span between 10.2 and 19.7 Ct/kWh.

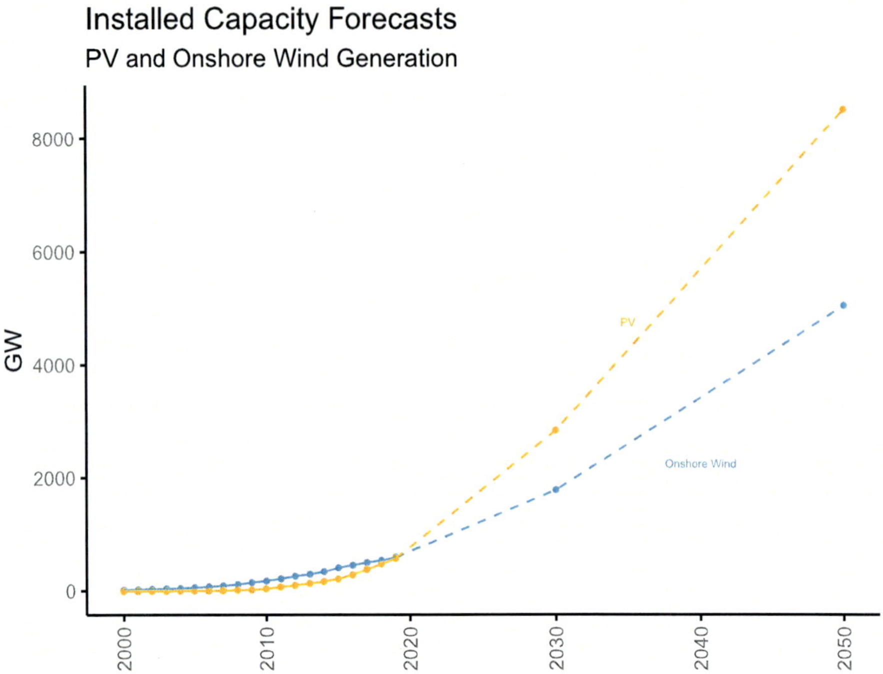

Fig. 3 Projected growth of PV and onshore wind power generation capacity

How does this development relate to Smart Grid developments? The increase in renewable capacity requires a different power grid from the historically established one. In the past, electricity was generated in large power plants by converting fossil primary energy sources such as coal or uranium with the aid of steam turbines. These large power plants were typically built near centers of consumption and were supplemented by run-of-river power plants on rivers, pumped storage power plants and gas-fired power plants. The energy needs to be transported via extra-high (380 k/220 kV) and high-voltage (110 kV) power lines and distributed to the consumers via medium-voltage (10/20 kV) and low-voltage grids (400 V). For this reason, the extra-high and high-voltage networks are referred to as transmission networks, while the medium- and low-voltage networks are referred to as distribution networks. On the transmission grid level in Europe, we typically find redundant system layouts following the n–1 criterion, meaning that security of supply is still guaranteed even if one critical component fails. On the medium-voltage grid, we observe both, redundant and non-redundant system layouts, whereas on the low-voltage level, typically a non-redundant layout is implemented. As described, great progress has been made in terms of technological efficiency and production costs of renewable energy generation plants (wind turbines and PV) over the last 20 years. Compared to coal-fired power plants, renewable energy plants are smaller and often installed on a decentralized basis. Solar PV systems, for example, are installed on rooftops of houses (and in recent years also in large PV parks on the ground), and thus, feed into the low-voltage grid. Wind turbines are usually installed at a certain distance from towns and villages or in large wind farms at sea. Onshore wind energy is typically fed into the low-voltage or medium-voltage grid. Both, the low-voltage and the medium-voltage grids, were not planned and installed taking feed-in electricity into account. In former times, these grid levels were exclusively demand sinks. Now, electricity is generated and fed into the grid on these levels. Unlike in the past, distribution grids are not pure energy sinks, but generation also takes place there which requires new technical solutions. The original network structures are not prepared for this, as load flow reversal can now occur, for example, i.e., electricity from the distribution network can flow into the higher voltage levels. This can be observed especially in distribution grids with a lot of installed PV and few consumers, which is the case, for example, in less densely populated rural regions where large roof areas (e.g., on barns) are available. Wind sites, on the other hand, are not necessarily located near the centers of consumption.

Controlling the grid on the low-voltage and on the medium-voltage level becomes necessary. For this reason, sensors and controllable network components are installed and turn the distribution network into what we call the Smart Grid. With an intelligent low-voltage grid, it becomes possible to manage supply and demand directly at this level. It is necessary to balance the distribution grid within the technical constraints (e.g., voltage band or frequency).

From an economic point of view, renewable energy plants should be installed where the highest yield can be achieved. This typically results in a regional distribution of wind turbines and solar PV depending on wind power and solar irradiation. Looking at the German situation, for example, we observe wind sites at the coast (North of Germany), whereas the centers of consumption are in the South which means that the solution lies not only in the Smart Grid but that transmission lines are also necessary.

In this book, we want to deal with this transition of the energy system and consider both technical and economic challenges and their solutions.

3 Smart Grids

The distribution network is not intelligent per se. Distribution networks provide the electric connection of end consumers to the power grid at the low-voltage electricity supply level. The low-voltage grid is connected to the medium-voltage network. At the transfer point, the local network stations, electricity is transformed from medium voltage to low voltage and then distributed to the end customers' network connection points. This results in a unidirectional flow of electricity.

With the increase of renewable power generation and additional electric load from new devices like electric vehicles (charging points) or heat pumps, the need to control the low-voltage distribution network increases: In order to feed electricity into the grid, the source must increase the voltage slightly above the grid voltage. If there is a lot of distributed generation connected to the distribution network, this can cause overvoltages or reversal of the electricity flow. Ideally, the decentrally generated energy is also used directly in the distribution grid to avoid transmission losses. Since the power generation of renewable energy sources depends on favorable conditions, a high degree of utilization is possible when the consumption side reacts to the power supply and, for example, cars are charged when excess power production exists. This required an intelligent energy management system that measures the status of both generation plants and consumers as well as the distribution network and intervenes with control technology. The control technology must therefore master the complexity resulting from the monitoring of the grid situation and the control of the plants on both the consumption and generation side. The aim is to maintain the technical grid conditions and to deal efficiently with load and generation peaks. In this way, the use of expensive reserve power can be avoided.

A prerequisite for this is that the network condition information can be recorded and processed. This also includes the status data of the generation plants and, in particular, of the consumers. Smart meters, i.e., electricity meters that record the power consumption up to the minute, are used to measure consumption at precise times. Time-based metering of generation and consumption not only creates the transparency required for control but also the possibility for further functions such as dynamic tariffs, fault monitoring, or the switching on and off of devices. Information and communication technology enables an intelligent and automated interaction of generation and consumption devices. This intelligent infrastructure is a fundamental prerequisite for new business models. Thus, the distribution network is developing into a smart grid. The term smart grid first appeared in the literature in the 2000s and has become more common since 2004 (Bollen 2011). However, there is no general agreed upon definition and worldwide different interpretations exist.

European Union

According to Jiménez (2011), the EU Smart Specialisation Platform[3] provides the following definition: "*A Smart Grid is an electricity network that can cost efficiently integrate the behaviour and actions of all users connected to it—generators, consumers and those that do both—in order to ensure economically efficient, sustainable power system with low losses and high levels of quality and security of supply and safety.*"

United Kingdom

The UK Department of Energy and Climate Change provides the following definition (The Smart Grid Forum 2014): "*A smart electricity grid that develops to support an*

[3] https://s3platform.jrc.ec.europa.eu/smart-grids.

efficient, timely transition to a low carbon economy to help the UK meet its carbon reduction targets, ensure energy security and wider energy goals while minimising costs to consumers. In modernising our energy system, the smart grid will underpin flexible, efficient networks and create jobs, innovation and growth to 2020 and beyond. It will empower and incentivize consumers to manage their demand, adopt new technologies and minimise costs to their benefit and that of the electricity system as a whole."

United States of America

The U.S. Departement of Energy provides the following definition[4]: "*[...], the digital technology that allows for two-way communication between the utility and its customers, and the sensing along the transmission lines is what makes the grid smart. Like the Internet, the Smart Grid will consist of controls, computers, automation, and new technologies and equipment working together, but in this case, these technologies will work with the electrical grid to respond digitally to our quickly changing electric demand.*"

China

According to Brunekreeft et al. (2015, p. 29) and Yu et al. (2012) the Chinese government has published a definition of smart grids: "*Smart grid technologies have the purpose to integrate new energy, materials and equipment as well as advanced technologies in information, automatic control and energy storage for realizing digital management, intelligent decision-making and interactive transaction in power generation, transmission, distribution, consumption and storage. Furthermore, smart grid assets optimize the resource allocation and satisfy diverse needs of customers as well as ensure the safety, reliability and cost-efficiency of power supply. Finally, the new technology [in the sense of smart technology] bridges the constraint of environmental protection and the development of the power market.*" The Chinese view includes an integration of provincial and regional grids with a strong (ultra) high voltage backbone.

International Energy Agency

A general definition, which is often used, was published by the International Energy Agency (IEA) (2011): "*An electricity network that uses digital and other advanced technologies to monitor and manage the transport of electricity from all generation sources to meet the varying electricity demands of end users. Smart grids co-ordinate*

[4] https://www.smartgrid.gov/the_smart_grid/smart_grid.html.

the needs and capabilities of all generators, grid operators, end users and electricity market stakeholders to operate all parts of the system as efficiently as possible, minimizing costs and environmental impacts while maximizing system reliability, resilience and stability."

In summary, all definitions have some aspects in common. Smart grids emerge as the union of different technologies: the physical energy network on the low voltage level, bidirectional communication capabilities for network participants, ICT and sensor technology to optimize and control bidirectional energy flows, and to enable new business models and services. The US definition explicitly points out the similarity to the Internet but underlines its application to the electrical grid. The Chinese definition differs from the others as it explicitly includes higher voltage level grids in its definition, while all other definitions make no specification in this respect. Typically, several goals are associated with the smart grid (Ekanayake et al. 2012, p. 6):

1. Ensure reliability and security of supply;
2. Optimize economically efficient grid monitoring and control, grid operation, and operation of assets (e.g., generation, controllable load, storage, grid technology);
3. Enable consumers to contribute to grid management;
4. Improve physical capacity and flexibility of the network;
5. Enable demand-side management and demand response;
6. Facilitate the integration of renewable energy generation into the grid;
7. Integrate communication, sensing, and measuring;
8. Resist attacks;
9. Enable and grant open access to markets and services.

In order to achieve these goals, a suitable technical architecture is required that enables interoperability between the various technical systems. In principle, a distinction can be made between different technical levels, which are structured hierarchically. Figure 4 shows the structure of the Smart Grid Architecture Model, which was developed by the Smart Grid Coordination Group as a standardization framework. The lowest level is the *component layer*, in which all physical components (system actors, applications, power system equipment, protection and tele-control devices, network infrastructure, computers) in the smart grid context are described. In the *communication layer* above, protocols and all mechanisms for inter-operational information exchange are defined based on the use case, function, or service and associated information objects or data models. This establishes syntactic interoperability and network interoperability. The communication layer enables the exchange of information objects that are exchanged between the functions, services or components. These information objects are defined in the *information layer*. One level higher, the *function layer* represents the functions and services derived from the business cases and their relationship. This creates the technical prerequisite for forming the *business layer* as the top layer, in which the regulatory and economic market structures, business models, products and services, and market participants are described. With the division into these levels, the previously described properties and objectives of the smart grid are technically structured and new business models are thus

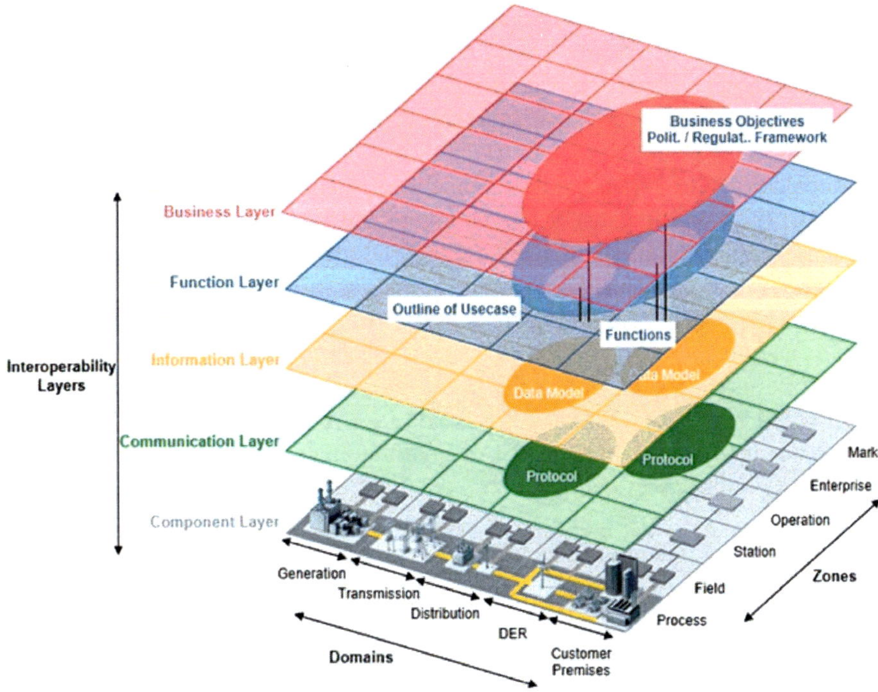

Fig. 4 SGAM framework (CEN-CENELEC-ETSI Smart Grid Coordination Group 2012, p. 30)

made possible. Accordingly, with the smart grid, not only technical challenges (e.g., efficient use of electricity in the distribution network, prevention of voltage band violations, control of load flow reversal, fluctuations in network frequency) become solvable, but new business models and services can emerge through the communication and information layer. For example, smart charging of electric vehicles can lead to capacity relief of the distribution grids or to better use of renewably generated electricity. Through appropriate control, renewable-generated electricity can be used directly in the neighborhood, e.g., through switchable loads such as heat pumps or stationary batteries. This gives rise to various economic and technical issues, which we will discuss in more detail in the further chapters of this book.

The transformation of the energy system to a carbon-free system requires major efforts and investments. This can only succeed if, in addition to the technical possibilities and what is economically feasible, there is also social acceptance. Therefore, the following guidelines for the transformation have been proposed by the Internet of Energy Working Group (Weidlich et al. 2013):

1. Subsidiarity: The distribution networks have local characteristics, which is why a generally applicable approach does not appear to make sense; instead, specific measures are necessary depending on the local framework conditions. Subsidiarity follows the idea of solving network problems first at the level where they arise and delegating them to higher levels for solution only if no solution is possible locally.

2. Flexibility as an economic good: as electrical energy requires permanent balance between generation and consumption, flexibility in generation, transmission, storage, and consumption becomes more important. Flexibility has an economic value. Thus, it can also be understood as a tradable economic good whose price adjusts to the respective situations.
3. Adequacy in infrastructure development and transformation: the distribution of costs, risks, and rewards in the course of the transition of the energy system should be appropriately distributed. Proportionality should be maintained and the interests of individuals and the general public should be balanced. This applies to all levels of the smart grid, from the infrastructure to the communication and information level to the market level.
4. Cost fairness: The costs of grid expansion and generation capacity expansions should include both external effects and take into account both pay-as-you-go components and benefit ratios. An undifferentiated apportionment to all consumers does not seem fair.
5. Incentives for Innovation and Investment: Since progress and efficiency result from innovation and investment, the necessary regulatory and technical conditions should be created so that, for example, grid operators have a choice between innovative solutions through smart grid technology and investments in grid capacity, rather than relying only on grid capacity expansion by default.

Even though great progress has been made in the field of the smart grid in recent years, the transition of the energy system is far from over. The smart grid is not yet the standard in distribution grids, although the technical need for widespread implementation is increasing. In the wake of the Paris Climate Agreement, the majority of countries are aiming for a zero- or low-carbon power supply. This means that the share of renewable generation units will continue to grow. At the same time, the transport sector is also undergoing a transition toward more electric vehicles, meaning that the number of charging stations will grow worldwide. In the next section, we will therefore look at the speed of the transition.

4 The Quest for the Social-Welfare-Optimal Speed of Smart Grid Technology Diffusion as an Important Element of a Sustainable Energy Transition

A transition of the electricity supply system toward exploitation of the benefits from smart grid technologies requires directed investments, social acceptance, and regulatory change. Diffusion of innovations, however, often takes a long time, sometimes decades (Rogers 1995). From a social welfare point of view, policy-makers should also take care that the promotion of innovative technologies is neither too fast (e.g., adoption of immature or overly expensive technology) nor too slow (e.g., forfeiting benefits in terms of competitiveness). Numerous important—and often interrelated—barriers exist, such as investment needs and financial resources; market uncertainty;

lack of a regulatory framework; low public awareness and engagement; lack of innovativeness in the industry; lack of infrastructure; technology immaturity; lack of the necessary technical skills and knowledge; integration of the grid with large-scale renewable generation; need for advanced bi-directional communication systems; lack of open standards; and cyber security and data privacy issues (Luthra et al. 2014).

Smart grid technologies are being implemented slowly, often only due to changed regulation (such as the mandatory installation of smart meters, which otherwise may take decades). Companies need to develop business models that create enough value to be worthwhile, for their customers and themselves, in a risky and rapidly changing environment. Today, because of the chicken–egg situation (lacking widespread availability of smart grid technologies versus lack of business models to make money), smart grid technologies are often not accompanied by business models, at a larger scale, that are able to exploit on smart grid functionalities.

The temporal change of a system's footprint is shaped by numerous different driving and other influencing factors. The dynamics and net cost–benefit ratio depend on the co-evolution of technical, societal, economic, and institutional partial systems. There are supply-push and demand-pull forces at work, and chicken–egg problems have to be overcome (e.g., electric vehicle sales are strongly dependent on the spread of the charging stations, design of standards, and financial policy support may be sub-optimally designed and crowd out private investments).

Welfare economics tells us that there is some optimal speed of transformation, determined by the expected costs and benefits which are the more uncertain the more distant one looks into the future. In the light of the enormous complexity of energy systems, and the social, economic, and ecological interrelations, such an analysis is challenging. Nevertheless, it will not be optimal to transform the system as fast as possible.

The power grid has evolved over many years, is a very capital-intensive and path-dependent infrastructure, and for many reasons constitutes a strong lock-in situation. An example is the long tradition of using alternating current technology for power transmission and distribution, and the largely still untapped potentials of direct current technology as an alternative.

Smart grids can have major impacts on consumers through (induced or required) changes in consumer behavior and lifestyles. Bigerna et al. (2016) have proposed a taxonomy of socio-economic aspects in terms of private (direct) costs, social costs arising from consumers' perceptions, privacy, cyber security, and regulation (Bigerna et al. 2016). Welfare gains and optimality—in a dynamically, potentially rapidly evolving system (where disruptive, game-changing 'smart' technologies are at play) with a lot of uncertainty—depends on resource availability, technological diffusion, political, and regulatory decisions but also societal valuation and acceptance.

The increased use of intermittent renewable energy sources, coupled with the increased use of distributed energy resources and the phase-out of conventional power generation, has positive effects in terms of decarbonization but negatively affects security of supply. To deal with the underlying challenges of this massive transformation of the energy system smart grids have to be developed. A range of smart grid technologies, including smart meters and advanced metering infrastructure, are being developed and gradually installed. Due to the rapidly increasing complexity of the system, and the largely unpredictable dynamics of this at the same time revolutionary and evolutionary process, system operators face an enormous challenge. An important question in this respect is how much can be left to the markets and how much needs to be centrally planned and controlled. The more long-term the perspective, and the more investment is needed, the more likely markets need to be directed. As an increasing part of the supply, demand and storage of energy, both temporally and spatially, is left to distributed resources and actors, the more dangerous it is to leave tasks to actors without responsibility for the energy system's overall performance and reliability. On the other hand, system operators cannot be expected to be able to orchestrate the entire system in all detail without being allowed to monitor the entire system in great detail as well, raising privacy and non-acceptance issues. Overall, there is a high potential for disruption, which in the case of more severe, and thus longer-lasting, electricity system outages, may have devastating consequences.

5 Policy Challenges

Decarbonization of the world economy necessitates also fundamental changes in the power system—i.e., a deep transformation of the electrical grid system, and therefore large (public or private) investments. Two major challenges are the integration of DER and electrification of the transportation sector. Both require a much more intelligent electricity network, i.e. smart grids (Cambini et al. 2016).

The transition from today's centralized, hierarchical and top-down organization of the electricity supply system towards a smart grid requires huge investments, new regulation and new ways of organizing (i.e., planning, constructing, and operating) the infrastructure. Smart grid investments are related to enhancing the power system infrastructure (e.g., measuring and sensor devices), the ICT infrastructure (hard- and software), and smart end-use devices.

However, in a smart grid framework there will also be totally new relations between the economic actors—not just suppliers and consumers of energy, but also prosumers. Overall, this transition can be expected to create major shifts in consumer behavior, lifestyles, and culture (Bigerna et al. 2016). Furthermore, it requires social acceptability and acceptance, which is affected, among other factors, by the perceived short- and long-term costs and benefits of smart grid investment. On the cost side, this involves not just investment and operating costs, but also costs of maintaining cyber security, costs of privacy loss, and others. It also requires new standards, regulation, and institutions.

The development of the smart grid infrastructure can be viewed as a co-evolutionary process of systems (technologies, institutions, user practices, ecosystems and business strategies) more or less aligned with energy (and climate) policies (Hall and Foxon 2014). For policy-makers it is crucial to better understand the drivers and processes of change, offering better chances to steer the various systems towards the envisaged sustainable energy transition.

The policy and governance challenge is to adequately support the co-evolution of systems, by mitigating non-monetary barriers and providing financial support and regulation to foster the transition. Given that the optimal level (and trajectory) of distributed versus centralized generation (and storage) is unknown, policy-makers should make an effort to define and operationalize the energy policy trilemma in order to find, and keep, a balance between the different trade-offs identified (e.g., social justice, security of supply, environmental sustainability) for the sustainable energy transition aimed for, and system transformation needed.

From a regulatory perspective, a main challenge is to decide how much can be left to the market and how much needs to be left in the hands of regulated entities such as DSOs to follow a predefined path of innovation and development.

6 Regulatory Challenges

Finally, from a regulatory perspective, smart energy systems, increasing energy system integration facilitated by ICT and machine learning (ML), as well as the re-emergence of multi-business utilities, calls for multi-sector regulatory bodies. These need to overlook the activities, define adequate regulatory frameworks (including penalties for non-system-compatible behavior), and to steer the course towards efficiency and system resilience. Efficiency is determined by efficient markets as well as economies of scale and scope (Jamasb and Llorca 2019, Sect. 3.3). In the literature, there is considerable debate on the pros and cons of single-sector versus multi-sector regulation. Needless to say that there is considerable unchartered territory out there with respect to how smart energy systems will evolve and diffuse society and the economy.

In Chap. 2, besides an outline of what distinguishes 'smart energy economics' from the more conventional energy economics, also addresses regulatory issues along the lines of the new paradigm "Energy system integration (ESI)", aiming at exploiting horizontal synergies and efficiency at all levels, based on sound economic principles and ex-ante and ex-post evaluation, as well as a well-balanced policy framework and a secure and resilient, integrated physical infrastructure, and enabling integration with other infrastructural sectors (e.g., transportation, telecommunication, water). The challenge is to quantify expected net gains of vertical separation (as it has been pursued for the last 30+ years) versus vertical integration (enabling integration of DER also across wider areas). Economies of scale need to be compared to economies of scope and of coordination (of inter-dependent activities, but also of jointly used knowledge—related to planning, policy and regulatory design etc.), the performance of multi-businesses needs to be gauged against the performance of unbundled market activities.

Chapter 5 explores some regulatory and institutional challenges related to smart grids, balancing basics of regulation with the topical developments around smart grid evolution. A fine balance needs to be sought in terms of competition versus coordination between networks (and network operators) and the network users. The three issues addressed are: (1) the adequate governance of (evolving) smart grids at the distribution level in order to enable non-discriminatory access to monopolistic networks; (2) appropriate regulation (cost-, price- or output-based) in the light of new tasks for network operators; and (3) network charging to address (decentralized) network congestion and designing well-functioning markets for flexibility.

Review Question

- What are the main drivers and challenges of the ongoing energy systems transition?
- Explain why electricity and power systems are so special for the sustainable energy transition.
- Describe what energy policy-makers need to take into consideration when implementing sustainable energy policy measures and programs.
- What are the main characteristics of smart grids?
- Describe the term 'sector coupling' and provide two examples.
- Name four misconceptions in policy-making related to the energy system transition.
- Summarize the main policy challenges related to the development of smart energy infrastructure viewed as a co-evolutionary process.

References

Agora (2017) Energiewende 2030: the big picture-Megatrends, Ziele, Strategien und eine 10-Punkte-Agenda für die zweite Phase der Energiewende. Technical report, Agora Energiewende, Berlin

Banschbach V (2003) Einflussgrössen des Energieverbrauchs-eine empirische Analyse für Deutschland. Master's thesis

Bigerna S, Bollino CA, Micheli S (2016) Socio-economic acceptability for smart grid development—a comprehensive review. J Clean Prod 131:399–409

Brunekreeft G, Luhmann T, Menz T, Müller S-U, Recknagel P (eds) (2015) Regulatory pathways for smart grid development in China. Springer, Vieweg

Cambini C, Meletiou A, Bompard E, Masera M (2016) Market and regulatory factors influencing smart-grid investment in Europe: evidence from pilot projects and implications for reform. Util Policy 40:36–47

CEN-CENELEC-ETSI Smart Grid Coordination Group (2012) Smart grid reference architecture. Technical report, Smart Grid Coordination Group, Document for the M/490 Mandate

Ekanayake J, Liyanage K, Wu J, Yokoyama A, Jennings N (2012) Smart grid-technology and applications. Wiley

European Union (2016) Paris Agreement COP21. https://ec.europa.eu/clima/policies/international/negotiations/paris_en. Accessed 24 July 2020

Hall S, Foxon TJ (2014) Values in the smart grid: the co-evolving political economy of smart distribution. Energy Policy 74:600–609

International Energy Agency (IEA) (2011) Technology roadmap smart grids. Technical report

IPCC (1990) Climate change. Cambridge University Press
IRENA (2019) Renewable power generation costs in 2018. Technical report, International Renewable Energy Agency, Abu Dhabi
IRENA (2020) Renewable energy statistics 2020. Technical report, International Renewable Energy Agency, Abu Dhabi
Jamasb T, Llorca M (2019) Energy systems integration: economics of a new paradigm. Econ Energy Environ Policy 8(2):7–28
Jiménez MS (2011) Standardization mandate to European standardisation organisations (ESOS) to support European smart grid deployment. Technical report, European Commission, Directorate-General for Energy, Directorate B—Security of supply, Energy markets & Networks, B.2—Electricity & Gas
Luthra S, Kumar S, Kharb R, Ansari MF, Shimmi S (2014) Adoption of smart grid technologies: an analysis of interactions among barriers. Renew Sustain Energy Rev 33:554–565
Math HJB (2011) The smart grid: adapting the power system to new challenges, volume Lecture #3 of Synthesis lectures on power electonics. Morgan & Claypool
Rogers EM (1995) Diffusion of innovations, 4th edn. Free Press, New York
Stern DS (2011) The role of energy in economic growth. Crawford School centre for climate economics & policy paper (3.10)
The Smart Grid Forum (2014) Smart grid vision and routemap. Technical report, UK Department of Energy & Climate Change
Walker C (2020) Using the United States wind turbine database to identify increasing turbine size, capacity and other development trends. Energy Power Eng 12:407–431
Weidlich A, van Dinther C, Briegel F (2013) Impulse für eine Smarte Energiewende—Handlungsempfehlungen für ein IKT-gestütztes Stromnetz der Zukunft. Technical report, AK Internet der Energie des Bundes der deutschen Industrie
Worldbank (2018) The energy progress report. Technical report, Washington DC
Yu Y, Yang J, Chen B (2012) The smart grids in China-a review. Energies 5:1321–1338

Smart Grid Economics

Reinhard Madlener

> **Learning Objectives**
>
> - Be able to understand the key energy economics aspects of the Energy Systems Integration paradigm,
> - To get a grasp on the system changes involved due to the use of smart grid (energy and ICT) technologies,
> - To be able to describe the types of new actors involved in smart energy systems and their roles,
> - To have a solid understanding of the major impacts on the economics of competing supply and demand-side flexibility options,
> - To obtain a sense for the potentials of the different kinds of distributed energy resources and their possible value for actors in smart grid systems, and
> - To be able to explain the role of machine learning in the context of energy economics and management of smart grids.

1 Introduction

The question why energy economics was established as a dedicated subdiscipline of economics is typically answered by pointing to the many specifies of energy markets, regulation and politics. Likewise, one could ask the question why there is a need for smart grid economics? Put differently, which tools in the toolbox of an energy economist are missing entirely, and which ones have to be adapted or otherwise modified?

R. Madlener (✉)
Chair of Energy Economics and Management, School of Business and Economics / E.ON Energy Research Center, Institute for Future Energy Consumer Needs and Behavior (FCN), RWTH Aachen University, Aachen, Germany
e-mail: RMadlener@eonerc.rwth-aachen.de

C. Dinther et al. (eds.), *Smart Grid Economics and Management*,
Lecture Notes in Energy 51, https://doi.org/10.1007/978-3-030-84286-4_2

Several of the smart grid topics addressed rather broadly are elaborated in some more detail in dedicated chapters, such as demand-side management (incl. dynamic pricing) (→ chapter "Demand Side Management"), energy market engineering (→ chapter "Market Engineering for the Smart Grid"), regulatory and institutional aspects (→ chapter "Regulatory and Institutional Aspects of Smart Grids"), system modeling (→ chapter "Modeling Smart Grid Systems"), (business) analytics (→ chapter "Smart Grid Analytics"), and business model design (→ chapter "Business Model Design"), and are therefore only discussed in a concise manner and mainly from an economics perspective.

The aim of this chapter is to point out not only some specialties of smart grid technologies and systems relevant from an energy economics perspective, but also the drivers of the paradigmatic changes ahead. These can be summarized as decarbonization, decentralization, digitalization, and democratization. While there are large potentials for efficiency gains, the challenges are formidable for the system operators for whom both distributed supply, demand, and storage/demand response will become less controllable and predictable overall, compared to the traditional top-down hierarchical system.

Smart grids enable to deliver energy in a much better controlled ("smart") manner, from different locations of generation to (active) consumers. Ideally, such a smart grid system is even more reliable and resilient than the classical hierarchical, top-down system (although this still needs to be proven; after all, the transition from system A to system B can be expected to be a lengthy process that is prone to many uncertainties).

Energy consumers today are increasingly empowered (and potentially also overwhelmed) by a growing number of choices including DER—such as rooftop solar PV, smart loads and energy storage systems, and energy service-oriented options, such as peer-to-peer (P2P) trading and participation in microgrids or some virtual power plant (VPP). Likewise, more cost-reflective tariffs aim at enhancing system efficiency and potentially offer cost-saving opportunities by acting flexibly and system-friendly (Gui and MacGill 2018). Such a continuously widening range of choices allows consumers self-selection in order to satisfy their individual energy (services) needs, affecting their energy consumption, production (prosumer households), and energy asset investment. Energy consumers/prosumers, however, have to make highly complex choices under considerable economic, technical/system, and social uncertainty. Still, to this end, a growing number of consumer-centric service innovations are available to assist energy consumers in their decision-making processes. And while some consumers/prosumers may prefer to get assistance only for individual services (for their personal, customized preferred solutions), others will prefer more fully integrated services (Gui and MacGill 2019).

Smart energy economics, in a way, combines standard energy economics with power system economics and elements of energy information science. It also brings together new business models and management concepts enabled by smart grids with analytical concepts that have been well established over the last decades—but somehow pushing it further to new levels. Examples of new analytical concepts are local market designs for peer-to-peer trading, the economics of automated energy management systems, and energy hubs, to name just a few.

The spheres or domains that can be distinguished are the following:

- Generation and storage,
- Transmission and distribution,
- Customer/end-use,
- (New) Service providers,
- Markets, and
- Regulation and policy-making.

Smart grids are part of an emerging paradigm change toward a more holistic, system-oriented perspective of the energy supply system. It will also require the development of new economic, regulatory, and policy frameworks to ensure efficiency. The question is out there whether the enabled sustainable energy transition and required transformation of the socio-economic and technical systems will be incremental or disruptive, but it makes sense in any case to discuss some of the issue at hand, to develop new analytical concepts and models, and to aim for proactive regulatory and policy guidance.

Different strands of the economics literature, not just neoclassical economics, are potentially useful for a better understanding of the issues and potential pathways and solutions at hand. For instance, we could think of 'new energy economics' (e.g., in the sense of making use also of new institutional economics, evolutionary economics, and behavioral economics) as a new school of thought, and in contrast to more conventional energy economics inspired mostly by neoclassical economic thinking (e.g., Zweifel et al. 2017). Hence, this chapter mainly points out the key characteristics of smart grids and related issues, as these are not typically covered in standard energy economics textbooks. It also draws on (often very recent) literature where aspects of smart grid economics and management have been discussed, typically from a very specific angle not easily accessible, and thus not well suited, as a reference for readers who want to effectively and efficiently build up their knowledge on smart grid economics.

The remainder of this chapter is organized as follows. In Sect. 2, we discuss objectives of smart grid economics as a new strand in energy economics, and *energy system integration* as a new paradigm, and also reflect on the potential benefits of multi-business utilities in a smart grid world. Section 3 takes the perspective of structures and actors, not only discussing the impact of smart grids along the value chain generation–transmission/distribution–consumer, but also discussing the role of new (energy) service providers, including aggregators, and the emergence of new markets. Section 4 tackles some governance, policy and broader issues related to smart grid economics, whereas Sect. 5 addresses the investment needs and related issues. Section 6 then briefly discusses the various time scales involved for economic analysis and value creation, including the need and value in short-term balancing. Finally, Sect. 7 focuses on the smart grid initiative in the United Kingdom as a case study.

2 Energy System Integration (ESI) and the Need for Smart Grid Economics

Smart grid technologies enable a restructuring of the energy systems as we know them in a way that has not been seen since the early days of electrification. Still, the co-evolution of the existing energy system/sector—with all innovations, standardizations, and regulatory adjustments—has taken many years. It can be expected that also the co-evolution toward a much more integrated, digitalized energy system will be a lengthy process (not least due to the massive investment needs and the fact that, at least modern societies, cannot afford a non-reliable energy supply system). But the new system does not only have to be reliable (and resilient), but it also needs to be affordable and environmentally sustainable (the energy sector, e.g., has been a major contributor to global greenhouse gas emissions).

The challenge of restructuring is all the more ambitious as it involves achieving or maintaining sustainability (along the economic, social, and ecological dimensions), energy justice, and supply security/resilience. It also implies that the inherent trade-offs (dilemmata) are actively managed and that the system remains adaptive (avoidance of undesirable lock-ins) and manageable (acknowledging the curse and limits of complexity).

2.1 Smart Grids and ESI as a New Paradigm

Smart grids enable system integration at an unprecedented level. Apart from new ways of coupling energy carriers and infrastructure systems, exploiting numerous horizontal synergies and raised efficiencies at all levels, in principle they will also allow to more effectively balance sustainability, energy security, and equity considerations. Apart from technical issues that still need to be resolved, there is an urgent need to think about the economic, regulatory, and policy frameworks that will enable to exploit these new potentials and to continuously safeguard the efficient performance of such SG-enabled integrated energy systems over time. In the following, a synopsis is given on the ESI paradigm (cf. Jamasb and Llorca 2019).

Aside from technical issues, there are needs to better explore not only the present and expected future economic and regulatory needs of smart grid systems but also unexplored business models and policy-making needed. The performance and desirability of a smart, integrated energy system will depend not only on the technical setup but also the economics behind it for both private actors and the public sector. In other words, there is no guarantee that such an integrated system will be superior unless it is well defined, designed, and orchestrated. Regulators are particularly challenged to steer the course broadly and avoid the worst consequences of a failure of a system that is many times more complex than the traditional ones.

Starting in the 1990s, with some earlier exceptions, market liberalization was becoming a dominant strategy to improve energy services through competition (not

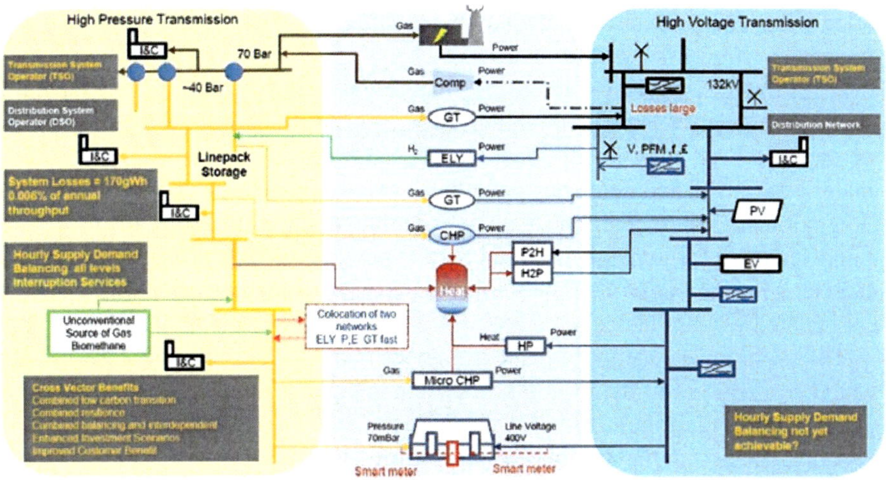

Fig. 1 Example of an integrated energy system architecture (Jamasb and Llorca 2019, p. 12)

only on the production/generation and retail side but also in terms of capacity usage of so-called 'natural monopolies'). ESI will challenge this existing paradigm by calling for the exploitation of synergies. ESI need more synchronization of the individual parts.

Figure 1 shows an exemplary architecture of an integrated energy system at the transmission and distribution levels, providing an idea of the enormous potential for flexibility and efficiency gains by means of potential substitution between energy sources to provide certain energy services to end-users. Note, however, that such benefits arising from a more integrated and thus more flexible system might come at higher capital and operating (in terms of managing) costs.

Such an integrated architecture can be seen as a 'network of networks' or 'system of systems' (cf. O'Malley et al. 2016), and an ESI may be extended to include also other network infrastructures (e.g., transport, telecom, water, and compressed air) than energy ones (electricity, heating, cooling, gases, etc.).

Since the 1990s, such network industries have been liberalized and unbundled (legally, or only in terms of accounting), in order to enhance competition but requiring effective and efficient regulation to avoid excessive profits and market power abuse. Obviously, the benefits of market competition and incentive regulation need to exceed the economies of coordination lost from ruling out vertical economies of scope. Traditionally, vertically integrated companies were able to benefit from horizontal economies of scale as well as economies of coordination stemming from vertical economies of scope (Jamasb and Llorca 2019, pp. 9, 14). Horizontal economies of scope typically arise from joint utilization of resources. Multi-utilities can benefit from providing services where the same network is used or where similar products, or services (e.g., billing), are provided jointly to the customers. In network industries, horizontal economies of scope can also be reaped through joint management of knowledge, e.g., regarding regulation, planning, R&D, and strategy/policy development.

As in many countries, market liberalization has led to some vertical separation of energy industries (so-called 'unbundling'), and there is obviously a dilemma: On the one hand, reducing institutional and regulatory barriers would help not only to unleash flexibility and efficiency potentials as well as synergies by means of vertical integration but also to integrate a much wider range of resources. On the other hand, it may lead to cross-subsidization and market power abuse, requiring new, and much more integrated models of regulation as well as to mitigate such inefficiency problems (and related welfare losses).

The evolutionary nature of the system development requires an equally dynamic system evaluation for identifying further/remaining coordination potentials and any (remaining or emerging) flaws. The quest is for a dynamically optimal and optimized system arising from the interplay of its constituent parts.

Overall, ESI enables the use of (technical, economic, social, etc.) synergies but requires some smart form of technical and economic/business coordination, both horizontally and vertically. Due to the energy system's increasingly distributed nature, it makes sense to organize it as much as possible based on well-functioning markets and transparently regulated and dynamic incentives that ideally not only ensure efficiency but also some degree of energy justice.

2.2 *Synergies and Multi-business Utilities*

An interesting question is whether the integrated energy company is outdated due to the neoliberal paradigm of introducing competition and low market entry barriers by unbundling. Still, capital markets might favor either more integrated or more focused utilities. The goal of multi-business enterprises is to create more value than can be achieved with stand-alone businesses. Both liberalization and re-regulation have been taking place simultaneously, and in different sectors of the energy market. The scope of synergy exploitation can be expected to have considerable influence on the value of business models in light of the further development and implementation of SG technology.

The challenges inherent in the need for massive structural change of the energy system in order to enable a sustainable energy transition toward a zero carbon energy supply also provide room for a multitude of new business models and opportunities. As discussed in Fuhrmann and Madlener (2020), synergies are subject to permanent change, thus requiring continuous evaluation of the corporate and social welfare surplus of multi-business utilities compared to focused utilities.

A classification of synergies is shown in Fig. 2, whereas Fig. 3 depicts the main cross-functional synergies between different asset classes (midstream/trading and others) of a multi-business utility. From Fig. 2, it becomes clear that there are many types of synergies (operational, managerial, financial, and so-called 'synergies of market power'). The operational ones have to do with costs or revenues and the

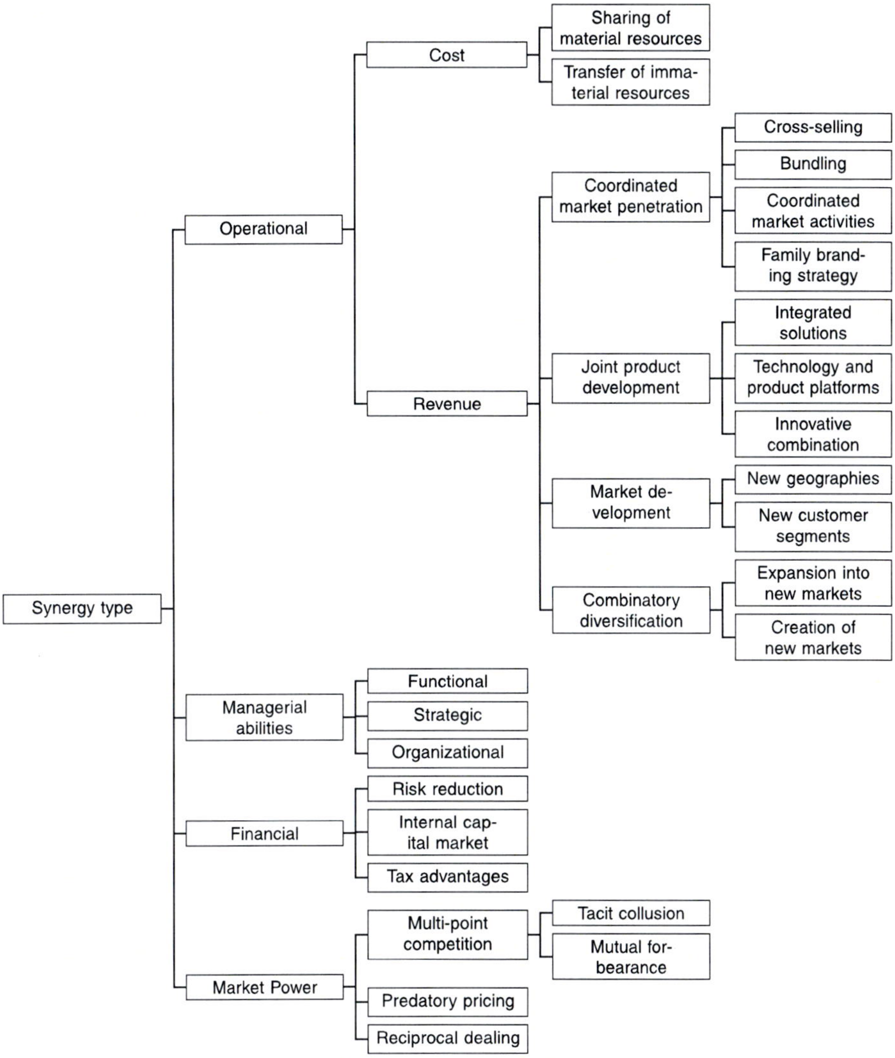

Fig. 2 Taxonomy of synergy types (Fuhrmann and Madlener 2020; Müller-Stewens and Brauer 2009, p. 376)

sharing or combining of resources. The managerial ones are related to functions, strategies, and organizational aspects. The financial synergies have to do with risk mitigation, company-internal capital market advantages, and tax benefits. Finally, the synergies of market power involve multi-point cooperation (taxit collusion, mutual forbearance), predatory pricing, and reciprocal dealing. A discussion of how to interpret these synergies in the context of a german multi-business energy utility is provided in Fuhrmann and Madlener (2020).

As can be seen in Fig. 3, which is adopted from that study as an example, all the asset classes are interrelated with trading, and within the trading business/stage each

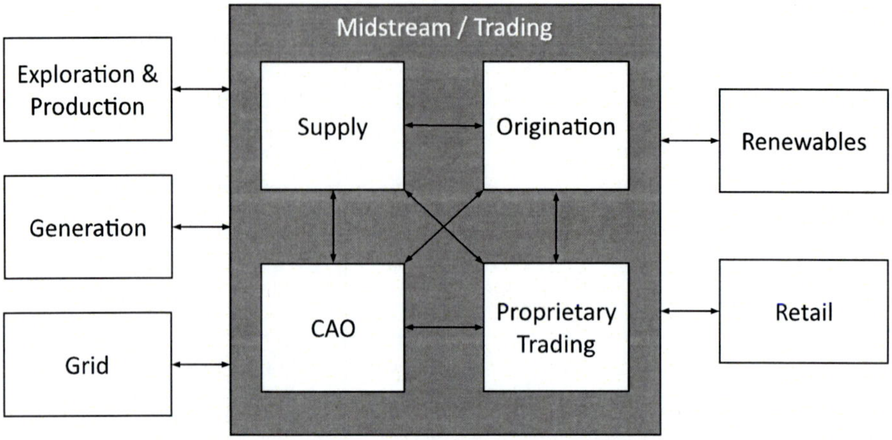

Fig. 3 Cross-functional synergies between classes of assets within midstream/trading and other asset classes (for larger utilities often supra-regional ones) (Fuhrmann and Madlener 2020, p. 29)

asset class has interdependencies with each other (mainly due to shared infrastructure and resources, such as cross-commodity trading desks or jointly used information services). A big advantage of exploiting synergies between the asset classes depicted can be reducing the cost of capital (e.g., due to internal value chains), and thereby increasing a company's liquidity.

3 Structures and Actors

Smart grids integrate energy and ICT networks, dedicated hardware and software, as well as numerous distributed (and centralized) assets that enable to monitor, control, and aggregate ('pool') generation, storage, and end-use assets (see also Fig. 1). The development of smart grids requires different expenditures in the energy supply industry (transmission & distribution grids, power plants) and in the end-user sector/s (industry, private households, transportation, etc.). Investment considerations are discussed in Sect. 5.

The smart grid/ESI paradigm involves the efficient exploitation of renewable energy, network automation, demand response, reactive power management, and much more. This increased use of smart grid applications enabled by smart grid infrastructure investments (for Europe until 2020 estimated to be in the order of 600 bn Euros between 2014 and 2035 alone; cf. Cambini et al. 2016), if well-designed and well-managed, enables to enhance system efficiency, resilience, and social welfare.

Integrating higher and higher shares of RES into the power system is a challenge in light of their intermittency. Smart grid technologies enable to exploit also many new sources of flexibility, which in sum will increase the elasticity of (flexibility) demand. It will require some accurate estimation of the available (i.e., individual and, if available, also aggregated) flexibilities on the supply and demand side of the market and the system operator's side. A flexibility merit order, as illustrated in Fig. 3, would take into account the demand and supply side of some flexibility market in order to determine the short- and long-term equilibrium in the flexibility market, and to find out the response behavior to any changes in the flexibility market price signal revealed.

Apart from scrutinizing the net economic gains that can be reaped, it is necessary to understand the major actors and types of systems and concepts involved. In the following, therefore, an overview is given of the structures and actors, their characteristics and functions, and some economic considerations, in light of the co-evolving smart grids.

3.1 *Operators/Owners of Distributed Generation and Storage Units*

Smart grids will enable small producers of energy (distributed generation, distributed storage unit operators) to participate in the energy distribution. Such local generation and distribution of electricity will change the low/er voltage level of the grid to an active layer with multi-directional power flows. Open, and ideally competitive, local energy markets, enabled by an ICT-upgraded ('smart') distribution grid, will enable to balance local supply and demand locally.

Up to now, even modern network-based energy supply systems (e.g., for electricity, natural gas, and district heating) are commonly designed and operated independently from each other. *(Smart) Energy Hubs (EHs)* as a potentially important element of future multi-vectoral (or integrated) energy systems provide an opportunity for system planners, operators, and also prosumers to decentrally couple, and technically and economically optimize, heat and power generation, conversion, and storage (Geidl et al. 2007).

Distributed Generation: Flexible Power Plants

In the future smart grid, it can be expected that many end-use devices and appliances will be connected in real time and obtain grid parameter values (voltage and frequency information). In addition, some supervising algorithms will be installed that serve to maintain resilience and prevent damages/outages by monitoring the real-world performance of the system against the control signals. This enables dynamic end-user involvement and can help to prevent any security of supply problems, e.g., in the case of management system failures as well as cyber-attacks or other unfriendly intrusion.

Figure 4 presents a conceptualization of increasing degrees of decentralization, viewed from a technical (decentralized vs. decentralized need for flexibility), managerial (central vs. zonal/local operation), and economic perspective (fixed/grid-wide grid tariffs and/or centralized markets vs. dynamic/local grid tariffs and/or local mar-

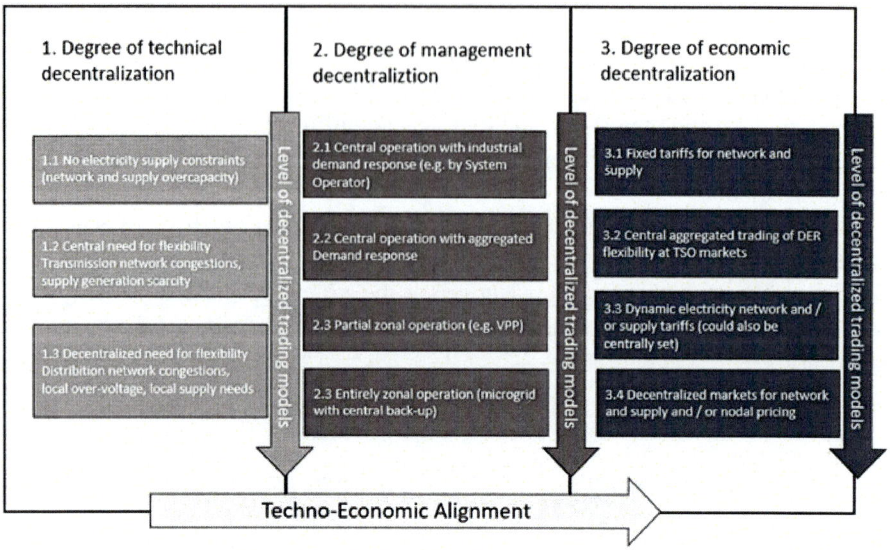

Fig. 4 Technical, managerial, and economic challenges with increasing levels of decentralization (Eid et al. 2016, p. 246)

kets). The transition could be from centralized large-scale markets for (aggregated) DER to an increasingly decentralized techno-economic management of the electricity system (Eid et al. 2016, p. 245).

The Economics of Storage

The potential economic value of storage has been discussed in the literature in recent years (e.g., Sioshansi et al. 2012; Crampes and Trochet 2019; Giulietti et al. 2018). Still, there are many storage technologies available, some of which can be installed in a centralized or decentralized manner, including such where some energy supplier or aggregator may not have access to (which reduces the potential, e.g., by trading with aggregated DER in the wholesale energy-only markets or some of the ancillary services markets). An essential question regarding the use of storage units in smart grids is whether there is enough economic incentive/value to operate profitably by, e.g., exploiting intra-day arbitrage opportunities (the time horizon of interest again depends on the technology). Storage technologies vary a lot in their relative characteristics and merits, but are typically characterized by their energy and power capacities, response times, ramping rates and per unit costs. Economics, regulation, market structure (and its development over time), and natural resources can vary a lot between countries as well, and so do the barriers for storage use and the creation of new business models for storage units, rendering generalizaions difficult (Madlener and Specht 2018).

The optimal level of storage use in a smart grid may change considerably over time, as the smart grid evolves, and it can be analyzed from an economic perspective either in terms of private and societal benefits and costs resulting from investment in and operation of a storage device.

Smart grid technologies in combination with storage units offer new and intriguing opportunities. However, regulatory and policy adjustments are necessary to enable the introduction of this important additional asset class in the supply or value chain (Zame et al. 2018, p. 1650).

3.2 Transmission and Distribution System Operators (DSOs)

A classical electricity supply chain categorization is generation–transmission–distribution. With the increasing integration of ICT into the electricity distribution system, and the change from a passive to a more active and 'smart' network, more system management is needed. As a consequence, the role of transmission system operators (TSOs) and DSOs will change. At the same time, the need for cooperation between the generators/suppliers of electricity and the DSOs and TSOs (and also between the latter) will increase; security of supply and system resilience will become a joint responsibility of DSOss and TSOs (Faerber et al. 2018).

The main investment burden toward a smart grid will be on the DSOs (Cambini et al. 2016). As regulated entities, DSOs will benefit from regulated tariffs/revenues to pass on to the grid users. While the path toward a full-fledged smart grid and integrated energy system can be smoothened by demonstration or pilot projects, still, in light of the tremendous change inherent in replacing and/or upgrading the established physical infrastructure by means of new sensors and controllers, and integrating more ICT infrastructure (and related software), the benefit–cost analysis of smart grid investments remains a big challenge (also in light of the fact that due to the need to decarbonize the energy system the timeline for accomplishment is less than 20 years). The uncertainty, aggravated by the time pressure, likely makes the smart grid much more costly than under more ideal planning and implimentation conditions.

With increasing shares of electricity from variable renewable energy sources (RES), system operators incur costs of integration. Hirth et al. (2015) classified these into three types: (1) grid-related ones; (2) balancing costs; (3) profile costs. The grid-related costs reflect the marginal costs (value) of electricity in space, referring to the opportunity costs of having to transport electricity from the place of generation to the place of consumption. Balancing costs arise due to forecast errors. Finally, profile costs, reflect the costs of matching demand and supply over time, and are the larger the more variable the intermittent output of RES is. Smart grids, enabling an efficient use of numerous flexibility potentials, can be expected to mitigate such RES integration costs.

3.3 Customers/End-users

Smart home and home automation technologies, containing a number of smart and integrated energy management components, also diffuse the market, enabling consumers to better optimize their energy use and match their needs, if they are prosumers, with their electricity generation and storage possibilities and preferences (Parag and Sovacool 2016). The increasing diffusion of smart grid technologies bears an unprecedented potential for interactions of prosumers with consumers, consumers with consumers, and both types of actors with a great many other actors. Energy consumer needs, behavior, and practices will be re-shaped and change, adding more risk and unknowns in the already complex transformation of the energy supply systems.

It can be expected that new customer classifications will emerge. Vulnerable customers in particular will have to be projected against rising energy bills, and may not be able to benefit from the incentives for prosumers, and may not be as responsive due to the inability to invest in smart home energy management systems and/or distributed energy resources. In order to avoid the 'death spiral' of grid operators suffering from rising numbers of self-sufficient (autarkic) end-users that defect from the grid, it might become necessary to use taxpayers' money to sustain the (smart) grids. Also, revenues from grid use tariffs that are mainly based on volumetric components can be expected to decline the less grid electricity the end-users (esp. prosumers) need.

3.4 Prosumers

'Prosumers' (a neologism formed of the two words 'pro-ducer' and 'con-sumer'; (cf. Toffler 1980) both self-produce and consume energy (sometimes also referred to as 'prosumage', cf. Green and Staffell 2017, to emphasize the storage unit as an integral component). The term is often used to describe a new role of private households in the politically pursued sustainable energy transition process. The role of energy prosumers is constantly evolving due to technological changes, leading to the market diffusion of new products and services, and creating new business opportunities as well as corresponding behavioral responses and demand changes, and maybe even lifestyles (Oberst and Madlener 2014). Prosumer households can be viewed either as individual, self-optimizing entities or, alternatively, as entities that are part of some energy sharing network enabled by smart grid technology, forming various kinds of communities (e.g., citizen energy communities, microgrids, and VPPs; see below) and allowing to address (environmental, economic, and social) sustainable development concerns. In the extreme, prosumers may have the ability to be autarkic/self-sustaining (provided the self-generation and storage systems are sufficiently large), and without interest to be part of an energy community, may even decide to opt for 'grid defection' (i.e., disconnecting from the grid), although this might be much more costly than 'load defection' (i.e., achieving net-zero balance between electricity taken from and fed into the grid) only (Sabadini and Madlener 2021).

Schill et al. (2017, p. 23) distinguish between the following four prosumage strategies:

1. *Pure prosumage:* Implies the complete avoidance of market transactions, restricting the optimization to the deferral of self-generated electricity to later periods via the storage unit;
2. *Grid consumption smoothing:* This implies that only prosumage storage loading from the market is enabled, allowing the storage unit to smoothen prosumers' electricity sourcing over both the own DER and/or the market;
3. *DER profiling:* This strategy involves the activation of only discharge to the market, enabling prosumers to profile their available DER feed-in (when it is most system-friendly);
4. *Full integration:* This implies no restrictions in terms of linking the DER with the market, enabling to use the storage unit for consumption smoothing, PV profiling, and arbitrage on some markets.

There is emerging literature trying to find out how consumer and prosumer households differ from each other (e.g., Oberst and Madlener 2014; Oberst et al. 2019). Also, there is evidence that renewable energy policy measures geared toward private households, such as the reduction of feed-in tariffs—aimed at making self-consumption of self-generated electricity more attractive compared to feeding into the grid—may actually increase electricity consumption (Atasoy et al. 2021). This raises interesting questions also for the implications of pushing the transition toward smart grid technologies forward.

Markets for 'prosumption' services are different from others—e.g., demand–response programs or platforms—in that prosumers can also offer active services to other prosumers, electric utilities, TSOs, and others have to bid for. Parag and Sovacool (2016) argue that prosumer marketplaces will be more complex if envisaged as a multi-agent system with very different types of services, a wider variety of participant groups/types fulfilling diverse and changing roles, and a larger number of providers per prosumption service. They distinguish between peer-to-peer models: (1) organically evolving peer-to-peer models; (2) prosumer-to-interconnected or 'island mode' microgrids; (3) organized prosumer groups.

3.5 *(New) Service Providers*

Energy Service Companies

Energy Service Companies (ESCOs) Companies have been well-established for many years, typically helping to save energy (energy costs). Still, it makes sense to review and reflect on their typical services and future potentials unfolding in light of smart grid systems and component management. In Chap. "Demand Side Management", their role is discussed regarding DR. energy service companys (ESCOs) comprise a wide range of market actors providing specific services which are typically not the task of DSOs. Through SGs, ESCOs are increasingly entering the household sphere, typically with services about energy monitoring data and better management of energy use (Verkade and Höffken 2018, p. 801). Hence, ESCOs can be seen as intermediaries between the household sphere and other energy system actors. An important subgroup among the ESCOs are the so-called 'aggregators'. They employ ICT to make use, and bundle, distributed flexibility, offering this flexibility as a service to grid management. Alternatively, it may sell bundled energy from local supplies on the wholesale market or green energy market. In business and industry, this kind of activity has been already quite common; at the household or neighborhood/city quarter level, it is still rather novel, and often has been restricted to the aggregation of distributed battery capacities.

In the following, we discuss how utilities/energy suppliers can enrich their business models by providing aggregation services.

DER aggregators may or may not be part of the utility business. System operators will benefit from a widened access to auxiliary services offered in the form of balancing services, voltage control (Lu et al. 2020).

Creation of New Business Models

The 'Energy Supplier 2.0', introduced by Specht and Madlener (2019), is a conceptual business model for energy suppliers aggregating flexible distributed assets. The focus of such new business models compatible with the ES2.0 idea is on the existing and possible new customer needs and market potentials. The concept, which is based on the business model canvas approach introduced in chapter "Business Model Design") embraces the notion of an aggregator of flexible capacities, e.g., on the household level, and unfolds how specific new energy business models can help to tap the potential of distributed flexible energy assets. Such assets can be seen as DR (see Chap. "Demand Side Management", but can also go beyond as well (e.g., including flexible generation and storage units as well)).

Smart grids, by enabling and stimulating customer/end-user empowerment and market entry of new players (e.g., ICT firms, 'big tech'), have disruptive potential also in the sense of 'creative destruction' in the energy business. An interesting question is whether electricity companies may either benefit or suffer from the deployment of smart grids—given that many of their established business models may no longer work (i.e., be no longer profitable). The main elements of any business model are *value creation, value delivery, and* value capturing (see also chapter "Business Model Design"). While many energy companies may eventually succeed to adapt and innovate their business model portfolios successfully, there is a lot of risk and uncertainty involved, e.g., related to customer engagement (needs and behavior), government support, and new market entrants/competitors. Several factors enable, whereas others inhibit, the transition of electricity firms toward doing business in the new, 'smart'

	Impact on Electricity Firms' Business Model Innovation: Enabling	Impact on Electricity Firms' Business Model Innovation: Constraining
Value creation	• New value propositions based on renewable energy & energy services • Increased demand for high-quality electricity • Customers' demand response helps balancing the network	• Empowered customers control and reduce electricity usage • Consumers generate their own electricity and become prosumers • Consumer engagement does not materialize
Value delivery	• Improved optimization of the electricity network • Improved marketing based on real-time data on electricity usage • Leverage assets of specialized ICT or energy service providers	• The value of conventional power plants erodes • The risk that new entrants become competitors • Increased complexity of the value network requires new capabilities
Value capture	• New revenue streams based on services and big data • Potential to become central actor in a multi-sided market • Reduced costs from less grid maintenance and load-shift	• Reduced revenues from selling electricity • Increased costs from investments in smart infrastructure • Uncertainty about the potential changes in revenues and costs

Fig. 5 Enabling and constraining impacts of smart grids on energy companies' business model innovation activities (in terms of value creation, value delivery, and value capturing) (Shomali and Pinkse 2016, p. 3839)

grid environment (Shomali and Pinkse 2016). The main impacts of smart grids on electricity companies' smart-grid-based value creation ability are summarized in Fig. 5, grouped by enabling and constraining factors and the three business model elements, *value creation*, *value delivery*, and *value capture*. It provides a balanced view on not only the chances but also the risks and uncertainties involved. The latter may incentivize firms to wait until major uncertainties have been resolved, or at least mitigated, and to scrutinize in detail whether complementary technologies, infrastructures, and institutions are developed in parallel in order to enable smart grids to actually come to fruition in the way it is hoped for.

Figure 6 depicts a set of companies active in business model innovation on the electricity retail market in Germany as well as a pattern grouping of different business activities (shown in a special type of 'energy trilemma' representation). It shows that most pattern groups investigated (top left corner) can directly link to a particular form of value creation. For example, the subscription and pay-per-use patterns are associated with mainly economic value creation, the open business, layer player and energy solution patterns mainly with environmental value creation, and cross-selling

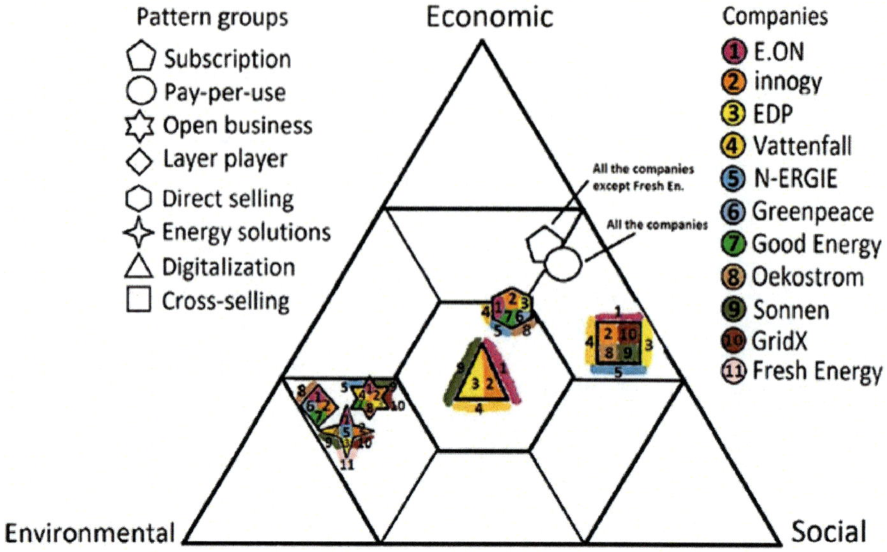

Fig. 6 Business activity clusters and companies active in smart-grid-enabled retail services in Germany (Karami and Madlener 2021, p. 13)

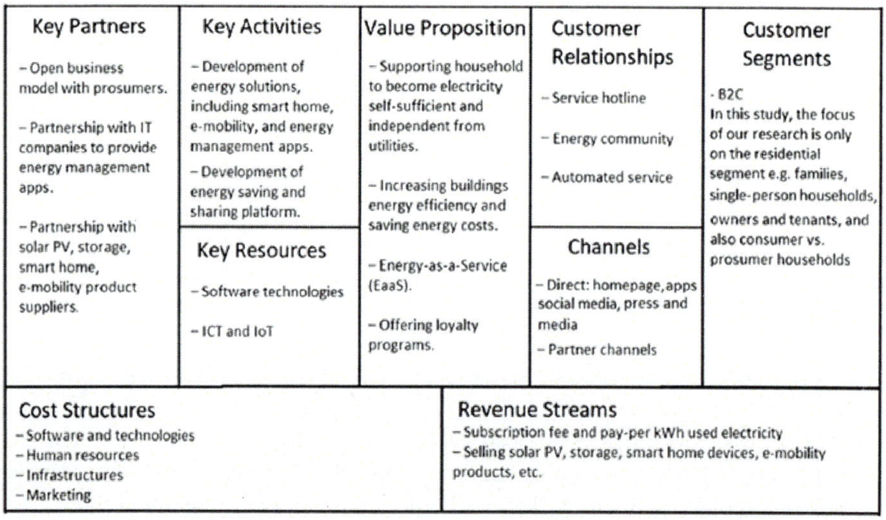

Fig. 7 Business model canvas for future electricity retailers (Karami and Madlener 2021, p. 13)

patterns mainly with socio-economic value creation. In contrast, the digitalization pattern is an integrative pattern group providing equal opportunities to all three forms of value creation. The same is true for the direct sale pattern group, although slightly less clear and somewhat biased toward economic value creation.

Figure 7 shows the proposed business model canvas (→ chapter "Business Model Design") for future retail electricity suppliers as the perceived optimal business model canvas. The intuition behind this proposition, which is corroborated by an empirical study of Karami and Madlener (2021) among 11 major German players in the retail electricity market, is that subscription and pay-per-use patterns for renewables electricity are found to be the most popular business models applied. By generating predictable revenues with lower sale costs, these patterns provide direct impacts on companies' revenue streams. Consumers presumably concentrate heavily on the high, and quickly rising, financial costs resulting from higher electricity end-use prices, along with the loss of time and effort needed to subscribe when deciding whether and what renewable electricity to buy. Such substantial transaction costs are seen as disincentives to subscribe, especially where tangible advantages are not straightforwardly apparent.

Citizen Energy Communities

Citizen energy communities (CEC), as the name implies, are geared toward local communities, with the aim that these are owned and/or (directly or indirectly) governed by the citizens themselves (the 'democratization' in the '4Ds', so to speak). Gui and MacGill (2018) identify three CEC types based on how communities interact with the energy system: *centralized*, *distributed*, and *decentralized* CECs (Fig. 8) . In general, CECs are "social and organizational structures formed to achieve specific goals of [their] members primarily in the cleaner energy production, consumption, supply, and distribution, although this may also extend to water, waste, transportation, and other local resources" (Gui and MacGill 2018, p. 95). This paper focuses only on electrical energy in the context of citizen energy. To illustrate this, in the following we examine the concept of different types of energy communities as well as potential benefits and challenges of citizen energy.

Centralized CEC are characterized by a relatively high level of cohesion, not necessarily in terms of spatial co-location but rather in terms of interaction. Its aim is to foster the achievement of common goals; members are normally directly connected with each other and conform to roles and social rules defined by the community. Rules and activities are typically managed by some governing body that is also controlling the communication and access to the members and with external parties (Gui and MacGill 2018, p. 100).

Distributed CEC are characterized as a "network of households and businesses that generate or own distributed generation individually, connected through a controlling entity either physically or virtually, and sharing the same rule in supplying and consuming electricity within the network" (Gui and MacGill 2018, p. 101). They are composed of a number of mostly homogeneous members who are not close to each other (in a spatial, normative, or cognitive sense). Most members are not directly connected with each other, and the boundary is transitory, partial, and permeable (linked by cross-cutting ties).

Decentralized CEC, finally, are "a community of households, businesses or a municipality that generates and consumes energy locally for self-sufficiency that may or may not connect to the main grid" (Gui and MacGill 2018,p. 102). The distinction from centralized and distributed CECs is due to its capacity for self-sufficiency and autonomy from the main grid. Decentralized CEC members typically belong to a spatially constrained area, such as a neighborhood, village/town, and municipality, and members may own DER individually, or collectively as a group, and sometimes even the distribution infrastructure as well (e.g., community-owned microgrid and integrated community energy system).

Figure 9 shows the citizen energy community setup in terms of actors involved and key relationships. A community-owned microgrid may involve different asset categories (generation, distribution, and microgrid), and be owned by single or multiple parties (e.g., the community, a utility, and other public/private enterprises). Generation assets include residential/commercial solar PV, storage units, and other DER (incl. demand-side ones). Distribution assets comprise all physical components that are part of the local distribution grid. Microgrid assets include the central controller of the microgrid, a central energy management system, smart meters, and a real-time communication and control unit.

Table 1 gives an overview of, and comparison between, the provider and customer relationships in a centralized electricity supply system versus one that is based on CEC (community microgrid). As can be seen, community microgrids are an emergent new customer service provider relationship which enables to reassess and redefine the role of customers. It allows for a new way of thinking about social inefficiencies and the disconnectedness of supply- and demand-side actors in the present-day

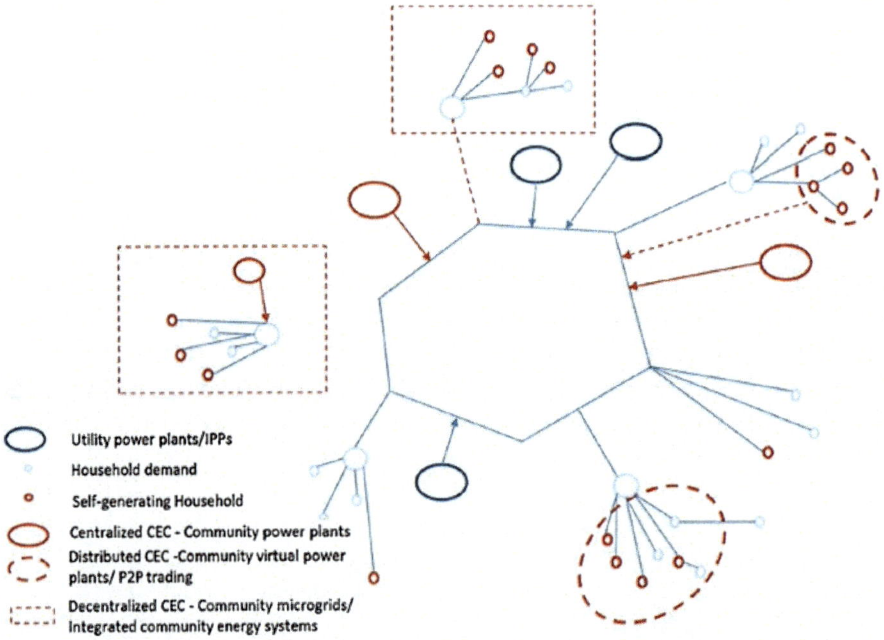

Fig. 8 Citizen energy community typology (centralized, decentralized, and distributed) (Gui and MacGill 2018, p. 100)

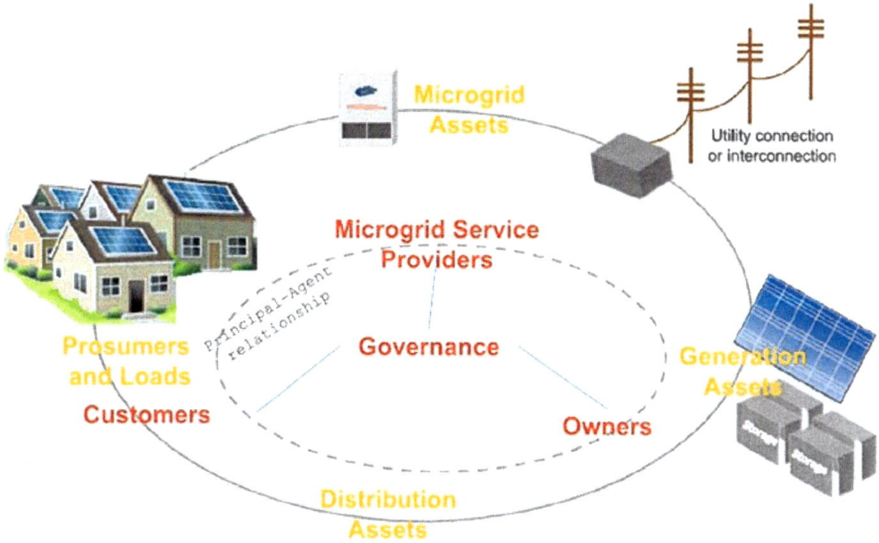

Fig. 9 Citizen energy community architecture (Gui et al. 2017, p. 1357)

Table 1 Customer and provider costs, benefits, and risks, centralized versus CEC power supply (Gui et al. 2017, p. 1361).

	Centralized supply	Community microgrid
Customer–provider relationship	Take-it or leave-it	Bilateral dependency
Customer involvement in governance	Low	High
Customer bargaining position	Individual	Collective
Investment cost recovery	Regulated user charges or market pricing	Ex-ante commitment (ownership or contracting)
Risk-bearing parties	End consumers for regulated services	Shared

centralized, hierarchical energy systems. Obviously, it also comes along with huge challenges in terms of structural, institutional, and regulatory changes to the centralized system required, and new business models (as discussed in the previous section, and in more detail in the chapter "Regulatory and Institutional Aspects of Smart Grids").

Economics of Microgrids, Virtual Power Plants, and Virtual Microgrids

Apart from the already discussed Citizen Energy Communities, many other ideas of how groups of consumers can be organized, and clustered, such that they benefit from grouping and following their joint interests and achieving their common targets, have been proposed for sharing the resources pooled, and potentially shared in a smart grid system.

Microgrids have been defined by the U.S. Department of Energy as "a group of interconnected loads and distributed energy resources within clearly defined electrical boundaries that acts as a single controllable entity with respect to the grid and that connects and disconnects from such grid to enable it to operate in both grid-connected or 'island' mode." The '4 Ds' point to an increase in the importance of microgrids in the future. Three sizes have been distinguished in the literature: small-sized microgrids (e.g., commercial buildings); medium-sized microgrids (e.g., communities); and large-sized microgrids (e.g., universities and military facilities) (Hanna et al. 2017, p. 47). Whether or not public benefits of microgrids are realized—e.g., in terms of improved power quality, grid reliability, improved black-start capability, lower cost of electricity or pollutant emissions—depends not only on the business model adopted but also the system boundary of the analysis. What might be beneficial for a specific local commercial building, a local community, or a university campus is not necessarily beneficial to society on a wider scale, as both in grid-connected or in island mode the microgrid might pose a burden on the public (i.e., the wider) grid. Likewise, there might be distributional effects resulting from the operation of the microgrid that are considered either as desirable or undesirable, depending on how the benefits and costs are shared between the energy end-users. Van Leeuwen et al. (2020) have investigated the social welfare implications of operating an integrated blockchain-based energy management platform for microgrid communities that also enables bilateral trading. For a real-world case study (a prosumer community in Amsterdam), they compare three scenarios (trade-only, grid-only, and hybrid) with a baseline scenario, finding that electricity import costs can be reduced by some 35% compared to the baseline, import volumes by 15%, and peak imports from the grid even by more than 50%. However, despite these impressive figures, total social welfare in the community was found to be highest without a trading mechanism, and the platform is only viable when sharing all costs equally between all households. Practical feasibility will also depend on social acceptance, and the social welfare might change over time, calling for some dynamic analysis or an evaluation over an extended period of time.

Prosumers can be clustered to virtual, orchestrated units, enabling the aggregator to participate in energy markets as one entity, thus greatly reducing total energy costs due to higher forecasting accuracy (especially if prosumers face penalties in the case of imbalances/load deviations) (cf. Vergados et al. 2016, p. 90).

The virtual power plant (VPP) concept has been widely investigated and used for managing geographically dispersed generation and/or storage units, typically managed by (or a utility or other energy company. Virtualization techniques and ICT are used for the optimized management (orchestration) of the DER (Vergados et al. 2016).

The VPP variant of Virtual Micro-Grids (VMGs), in contrast, has slightly different characteristics (cf. Vergados et al. 2016, p. 92):

- Management of very small energy prosumers by an aggregator (in contrast to large generation/storage units owned by a large utility company);
- Any type of small-scale facility able to produce or consume a small, or negligible, amount of energy (e.g., some public/municipal lamp-posts);
- Very different monitoring and control functionalities (e.g., low-cost VINSEM gateways that are customizable, backward-compatible, and communication protocol-agnostic);
- The VINSEM solution proposed is offered to the VINSEM prosumer who needs to install the VGW and required sensors and control equipment. In contrast to standard VPP cases, the VGW adopts some open-source software implementation, thus not requiring complex smart grid standards;
- VMG system requirements are different (e.g., data acquisition, communication, decision-making, and active VMG control management);
- Only RES prosumers are considered (thus aggravating the challenges due to RES intermittency);
- In contrast to VPP concepts, where one actor (the DER aggregator) manages the assets, the VMG concept foresees a new actor (the VMGA) which cooperates with various SG stakeholders (DSO, TSO, balance responsible party (BRP)) to enable the VMGA to participate in the (liberalized) electricity markets, and to react to specific events (e.g., in the case of a local congestion problem declared by a DSO, the VMGA needs to be sure that only prosumers associated with a particular low-voltage substation are eligible to participate in the prosumer clustering process).

3.6 *Local/Energy Markets*

The deployment of DER enables to turn regular (passive) final consumers into active contributors to the local supply of electricity, both in terms of energy, capacity, and reserve/balancing energy. Digitalization of the power distribution grids ('smart grids') and innovative regulation enables peer-to-peer trading, (as, e.g., the last amendment of the German EEG contains; cf. EEG, 2021) but the design of local energy markets is still in its infancy and demonstration projects still prevail.

The increased use of variable renewable energy sources raises the need for flexibility that enables it to respond quickly to fluctuations in supply and demand. All the different flexibility options at large- and small-scale (e.g., flexible generation, demand response, and storage), and particularly those enabled by smart grid technologies, could be thought of competing in a dedicated new market for flexibility where flexibility providers (typically aggregators) and other parties in need for flexibility (typically grid operators) meet and trade with each other on a level playing field (Council for European Energy Regulators 2019). Such markets can be thought of operating locally, regionally, or nationwide, potentially enabling to reduce losses due to efficient, low-cost local load balancing. Multi-layer trading of flexibility on dedicated platforms, and increasingly automated decisions of smart DER enabled by ICT, will allow an increasingly effective and efficient orchestration of the manifold resources playing an active role in the energy system. Transparency and a clear regulatory framework for DER will be paramount for efficient, and ideally also social welfare-optimal, flexibility market outcomes. In the following, some more issues related to local energy market design and peer-to-peer trading are discussed.

Local Energy Markets and Peer-to-Peer Trading

In this section, it is described how an auction model for a local reserve energy market can be designed that enables to also accommodate the special needs of non-expert bidders such as private households. The model can be used to revolutionize the reserve energy market, as a glsbrp, in contrast to today's standard practice, and is given the chance to self-supply reserve energy. Thereby, it serves at least two purposes: (1) it helps to further integrate DER; (2) it can help to lower the costs for reserve energy by mitigating the market power of the currently dominating, large-scale utility companies.

End-use energy consumers can benefit from this newly designed market twice, in that they are the ones providing the energy and getting paid for it, and in that their energy bill can be lower once the market offers reserve energy at a lower cost. At the same time, the mechanism supports the remuneration and subsidy schemes for DER that are already in place. In the longer term, when subsidy schemes are eventually phased out, it can serve as a long-lasting incentive scheme for investments in the designated technologies. It can be shown theoretically (Rosen and Madlener 2013), for a symmetric and an asymmetric setup, that the information policy in the market has a significant influence on the speed of convergence and also a small effect on the equilibrium market price that is finally reached. In the extreme case where no information is provided, the effect on the equilibrium price becomes substantial. Even more importantly, this effect is sustained indefinitely, which points out the importance of the market design choice.

Given the special characteristics of bidders in a local energy auction, a problem that needs to be solved is to find an adequate and at the same time reliable remuneration for each provider of reserve capacity and energy. Such an auction mechanism needs to be simple and easy to understand in order not to turn off potential participants. Furthermore, transaction costs in a market with such small quantities need to be low in order to leave room for at least a minimal profit, and opportunities for strategic behavior should be kept to a minimum. The design of an auction for such a purpose comprises many parts. Auctions for electricity are a specific type of auction, as the good is perfectly divisible and non-storable, which means transactions need to happen in real time, or at least at a predefined point of time in the future. The type of auction required can be compared to the treasury auction, which has received considerable scientific attention in the past. So far, game-theoretical analysis of reserve energy auctions with the properties needed in a local market is still very limited.

P2P trading enables direct interaction between local market participants without the involvement of third parties. As such, it is an alternative that also enables to switch energy suppliers on a high frequency (e.g., minute-by-minute) basis and to buy and

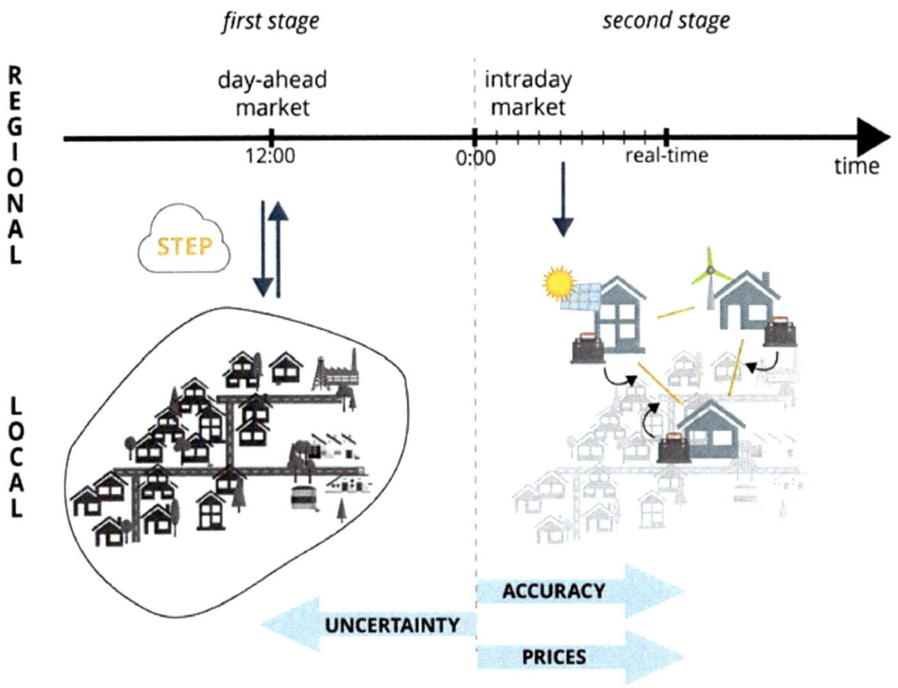

Fig. 10 Peer-to-peer trading platform architecture (Zepter et al. 2019, p. 165)

sell electricity based on one's individual preferences (prosumers as active 'producer-consumers'). P2P might involve blockchain technologies in order to keep track of the transactions made and to offer a transparent and at the same time automated settlement of the market transactions that occurred (van Leeuwen et al. 2020; Morstyn et al. 2018).

Still, such P2P energy trade concepts are still at a very early stage, and there is a lack of consensus regarding what market design or business model is best to develop such advanced local energy markets, and how the interplay with the established electricity markets (intraday, day-ahead) is to be organized (Lüth et al. 2018). It leads to several market design questions, some of which are discussed in the chapter "Regulatory and Institutional Aspects of Smart Grids" of this textbook.

Crucial issues are the merit order of the various flexibility options over time, the fact that these may, or may not, all be offered in a single market for flexibility (which would bring them into direct competition and lead to some cost-efficient outcome), and the many interdependencies that not only lead to high uncertainties regarding investment decisions (potential 'missing-money problems', e.g., if a certain business model suddenly becomes obsolete and unprofitable) but also energy end-user costs (esp. small-scale energy consumers are often risk-averse and do neither have the expertise nor the capacity for hedging their risks, which will require aggregators and other players to offer such services).

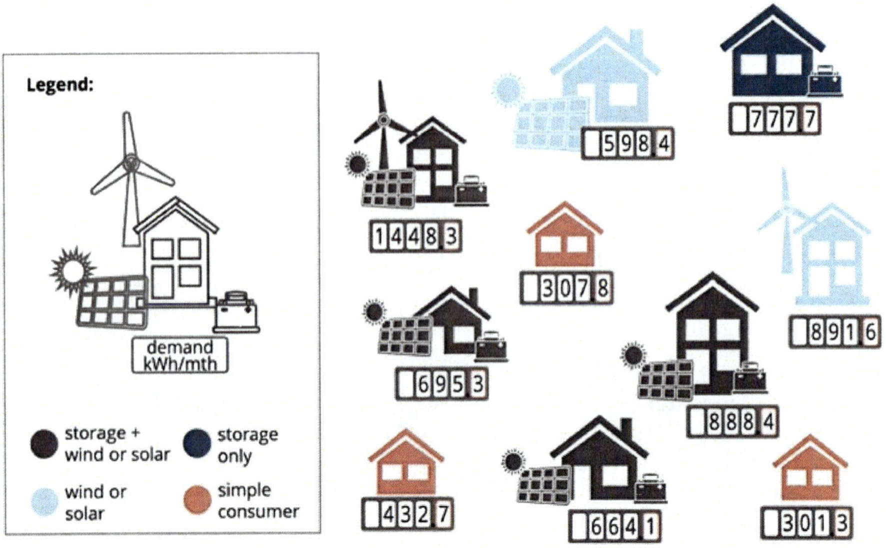

Fig. 11 P2P community and diversification synergies (Zepter et al. 2019, p. 169)

Zepter et al. (2019) study how prosumers can be integrated into wholesale electricity markets, and what synergies can be reaped by P2P trade and the use of residential storage units. Figure 10 provides an overview of how the system architecture looks like. The prosumer community is depicted as a number of buildings/households; based on a renewable power generation forecast, these bid into a day-ahead market. At this stage, each household plans its grid consumption, P2P trade, battery storage utilization, and the grid feed-in. The community submits at this stage a commitment to the day-ahead market that takes into account the wholesale electricity price and local wind and solar power generation uncertainties. In a second stage, the community needs to balance any deviations from the day-ahead market commitment by adjusting P2P trade, battery usages, and grid electricity procurement from the intraday electricity market. The households are somewhat heterogeneous (enabling some diversification effect), and are assumed to be connected both to the main grid as well as interconnected with each other by a local grid. The objective of the optimization problem could then be to minimize the community's expected costs of procuring electricity from the (intraday and day-ahead) wholesale markets.

The interplay and synergies of pooled heterogeneous households engaged in P2P (as well as wholesale market) trading are illustrated in Fig. 11. It visualizes the model community constructed, and shows the basic characteristics of the buildings/households considered in terms of technology portfolio and load.

In a study on the value of local P2P trade, Lüth et al. (2018) investigate two different market designs in the context of battery flexibility (decentralized, privately owned batteries in private households versus a centralized, commonly accessible, and thus shared battery). They find that P2P trade can save more than 30% of the electricity costs for a community, allowing for a significant increase in self-sufficiency, and utilization of local renewable energy resources. Note, however, that in such studies the assumption is made that the smart grid and related digitalization technologies are already installed (i.e., not accounting for those costs, considered as 'sunk costs', and financial risks).

4 Governance, Policy-Making, and Broader Regulatory Implications

Smart grid technologies enable the adaptation of the electricity supply system to the challenges ahead. The development path requires a coherent policy and regulatory framework that enables, and safeguards, a smooth innovation path. Apart from economic, social, and ecological sustainability, a major energy policy goal is the security of supply, especially when it comes to the power supply system which is particularly complex and sensitive to supply-demand shocks. The latter is increasingly investigated under the heading "resilience".

A major challenge from a regulation perspective is to find out, and then to decide, how much business can safely be left to the market (with an adequate framework that safeguards a level playing field) and how much needs to be left in the hands of regulated entities such as DSOs. In any case, the regulator needs to be ready and running to deal with the challenges ahead.

Future business models and cases depend on the ability to provide multiple services—such as reserve energy and capacity, balancing energy, and arbitrage. The economics depend strongly on the regulatory framework. The regulatory framework in place does not match the rapidly increasing complexity of smart grid energy supply systems. Researchers, regulators, and policy-makers alike are asked to proactively reform markets, regulation, and policies toward smarter energy systems.

Proactive regulatory change and innovation can help not to slow down the evolution of new (utility) business models. Grid users and new businesses will arbitrage the widening gap between new technological and market realities and the established regulation. If regulatory change cannot keep up with the changes happening to the energy supply systems, and the electric power system in particular, then large inefficiencies might occur (MIT Energy Initiative 2013).

Smart energy systems, increasing energy system integration facilitated by ICT and ML, as well as the re-emergence of multi-business utilities call for multi-sector regulatory bodies that will have to closely overlook the activities, define adequate regulatory frameworks (incl. penalties for non-system-compatible behavior), and to steer the course toward efficiency and system resilience. Efficiency is ultimately determined by not only efficient markets but also efficient and effective institutions and proactive regulation, as well as the exploitation of economies of scale and economies of scope (Jamasb and Llorca 2019, Sect. 3.3).

4.1 DER Regulation

DER can provide multiple services, depending on the type of DER (storage being particularly versatile), including energy arbitrage, investment deferral of conventional (generation, transmission, and distribution) capacity, ancillary services, ramping, end-user applications, and curtailment of power generation from renewables (for a discussion specifically for energy storage as a specific type of DER, see Sioshansi et al. 2012, p. 48). Still, today there is a situation of incomplete markets and thus also incomplete quantification, capturing, and valuation of such services that often stem from multiple value streams. As a consequence, there are also no incentives in place yet for siting DER at locations where they provide the greatest benefit to the system. The often still lacking market diffusion of smart grid technologies precludes the real-time dispatch of DER and the provision of numerous services that will eventually be possible once the smart grid has become a widespread reality. The higher complexity of DER integration from the system planner's perspective, and the lack of a more holistic simulation of networks with DER by the system operator/s can be expected to slow down, or even prevent, part of the DER installations and exploitation for many more years, even in cases where they would be expected to be more (cost-)effective than utility- or DSO-owned assets controlled by them primarily in terms of system support. Investors in DER can be thought of being given the choice to either sell their flexibility freely in some emergent flexibility market, or to benefit from some clearly pre-specified regulated service offered under some regulatory regime and rate base—as it has been common for conventional power generation, transmission, and distribution assets. Not only from a social welfare but also a business perspective, it is important to find out which regime is preferable (from a static and dynamic viewpoint, or in the shorter and longer terms, respectively). The lumpiness and irreversibility of investments, but also the presence of market power, can lead to inefficient and suboptimal investment in DER, although lumpiness and irreversibility can be assumed to be the smaller the more decentralized (and modular) the assets in question are. Modeling and understanding possible strategic behavior—of all actors involved (incl. DER operators) is important to understand the private and external (or systemic) value of DER. Both expected future costs and revenues from a DER investment are hard to quantify, partly by the continually changing environment, the need to re-optimize the DER operation, and uncertainty regarding flexibility market prices over multiple time scales (Sioshansi et al. 2012, pp. 52–53).

4.2 Smart Grid Regulation

Policy-makers are challenged to take the right governance decisions in terms of regulating the markets involved in the smart grid development—including, standardization and standards, DER, prevention of cyber-attacks and privacy infringements, system resiliency, and sustainable development. This enables them to adopt and adapt their support schemes in order to mitigate investors' risks and to enable them to properly value and capture the benefits and incentivize system-friendly behavior allowing for social welfare maximization overall. In light of the enormous complexity of not just the co-evolution of the smart grid but also all the services enabled now and in the future, such a holistic social welfare cost–benefit analysis is a formidable task to accomplish (and to track over time in order to allow some timely ex-ante, or else short-run ex-post, corrections of any shortcomings that may arise).

4.3 Governance and Policy-Making

New governance and regulatory policies are needed to shape the evolution of smart grid-based energy systems, and to enable the distributed and flexible assets to be orchestrated for a flexible and resilient supply system that is able to efficiently deal with all sorts of dynamics that might occur, and without jeopardizing security of supply. Ideally, it is also able to maintain, or even enhance, both a certain level of competition amongst suppliers of resources as well as of distributional justice (to be discussed amongst society what this means). Also, institutional inertia have to be tackled, new knowledge built up to be able to design new policies and regulation as the system/s evolve/s (e.g. on consumer engagement, cyber- and physical security, resilience in different dimensions). Ideally, governance, policy and regulation stays

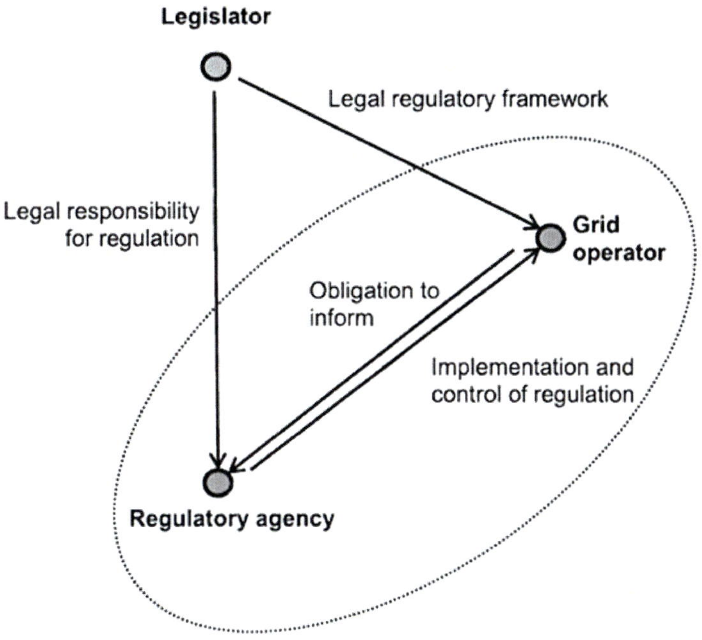

Fig. 12 Participants in the (smart) grid regulation process (Nolting et al. 2019, p. 756)

abreast of the changes, in order to enable ex-ante measures, rather than reacting ex-post (i.e. in retrospect), which likely creates higher costs than if measures are adopted on time and with some foresight about expected trends. Also, regulatory guidelines to promote the development of the smart grid should aim at improving cost efficiency of the power (or further integrated) system, but also try to optimize social welfare, embracing as many aspects as possible and manageable (e.g., distributed generation, network automation, demand response, reactive power management; cf. e.g., De Oliveira-De Jesus and Henggeler Antunes 2018).

An important barrier to smart grid investments is the lack of an appropriate regulatory framework (see also the chapter "Regulatory and Institutional Aspects of Smart Grids"). Besides, timely investments in electric grids can (and need to) be adequately incentivized.

Nolting et al. (2019) study the proper design of incentives for grid infrastructure investments based on social welfare considerations, and taking information asymmetry (principal-agent problems) into account. They further propose a so-called capital expenditure adjustment rule aimed at avoiding delayed cost recognition and leading to ill-incentivized (delayed) investment. It remains to be seen how useful the approach is under high uncertainty, as in the case of smart grid investments, where the value of waiting might severely delay the investment, which however might be perfectly rational (and even social welfare optimal, which needs to be assessed case by case). Principal–agent problems, (the lack of) property rights, and information asymmetries need to be taken care of, too. Figure 12 depicts the main participants engaged in the grid regulation process. As can be seen, three participating parties interact with, and mutually influence, each other in the grid regulation process:(1) the legislator; (2) the regulatory authority; (3) the regulated DSOs. Due to the information asymmetries between these participants, different principal–agent relationships may occur. For instance, conflicts between the goals of the regulatory authority (as the principal) and the DSOs (as the agents) may arise. While the regulator's target is to raise/maximize social welfare (by reducing total energy system costs, etc.), the DSOs aim at profit maximization. The problem due to diverting goals can be mitigated, e.g., if the incentive regulation aims at inciting the DSOs to act in a more efficient and welfare-oriented way—by linking their profits to societal goals. Information may also be asymmetrically distributed between the parties concerned (e.g., in terms of an information deficit on the side of the regulatory agency with regard to specific operation and management data of the individual DSO). Still, the regulatory authority may have some information advantages, too, e.g., by having an overview of the whole regulated sector, whereas individual DSOs usually only know their own data.

5 Smart Grid Investments

Smart infrastructure technology and information system investments are needed to cope with changes in demand (patterns and levels), as well as increasing shares of variable power generation. New pricing schemes and mechanisms will be necessary for funding both the investment costs and operation & maintenance (O&M) costs, as will be re-regulation that takes the change of the technical standards and the evolution of the energy supply system into account (on regulatory issues, see also the chapter "Regulatory and Institutional Aspects of Smart Grids", on the analysis of evolution strategies based on synthetic grid topologies see Pagani and Aiello 2015).

Values that accrue from smart grid investment can arise from different business models (and to different stakeholders involved). If only costs and benefits of individual smart grid investments are considered that accrue to the investing actor (e.g., a utility), then further value creation is not accounted for. This calls for an extension of the boundary for value (or cost–benefit) analysis arising from smart grid investment.

In principle, the cost for enhancing the current distribution grid toward a smart grid needs to be gauged against the benefits of higher connectivity. The main grid cost components are (1) losses (both in lines and at transformer substations); (2) security and capacity factors (robustness—e.g., evaluated by random removal strategies); (3) line redundancy (e.g., by analyzing a random sample of the nodes in the network and calculating the first n shortest paths of increasing length; the worst-case path between two nodes is considered); (4) power transfer limits (e.g., in terms of maximal and average current supported by a line) (cf. Pagani and Aiello 2015, p. 170)).

Investments in smart grids and smart grid applications need to pay off—both from an individual actor (private households, firms, and utilities) and social welfare point of view. From a distribution grid operator's point of view, it could be analyzed how long it takes to amortize smart grid investments when a locational–marginal pricing scheme is applied (say, due to the regulatory framework in place). Such research has been done in the past, either using relatively simple economic models or applying models that take the multi-level structure of smart grids explicitly into account. Some of this research also deals with social welfare implications and social welfare optimization. An example of this kind of research is Jesus and Henggeler Antunes (2018), where the authors have explicitly analyzed the benefits of DG, network automation, demand response, and reactive power management, respectively, exploring progressive levels of implementation (in terms of expenditures in smart grid technologies). In their analysis, they find that in weak power systems, DR applications are suitable to be recovered in the short term through the marginal revenues gained by the network provider.

On the one hand, ideally, increased flexibility from the various flexibility options will reduce the need for infrastructure investments. On the other hand, major new investments are needed to make the system smarter.

While the potentials for efficiency gains through smart grids are largely undisputed, the issue of investment and related risks and uncertainties is more controversial, and needs some more attention. Investments in smart grid infrastructure are a key enabler for reaping the benefits of smart grid systems, as well as for exploiting the net gains of 'smart investments'. Still, there can be a mismatch between where benefits and costs occur, leading to problems of value capture and redeployment (Hall and Foxon 2014). Also, some smart grid benefits are harder to quantify, and to price directly, such as security of energy supply and decarbonization (both can be considered as *public goods*). Value capturing and structural incentives need to be well understood, also in a (co-)evolving system—i.e., dynamically over time. Hall and Foxon (2014) discuss the political economy of smart distribution systems. Drawing on urban political economy they discuss in detail how smart grid investments can benefit municipal economic development, and are altered by changes in municipal-level value creation. In addition, they also characterize the co-evoluation of value capturing and structural incentives in the power distribution system. Obviously, the benefit of smart grid investments will also depend on consumer acceptance and willingness to pay for smart services (Lineweber 2011; Toft et al. 2014).

Apart from social acceptance, and 'socio-economic acceptability' (Bigerna et al. 2016), there can be numerous barriers, and interactions among them, hindering the adoption of smart grid technologies (Luthra et al. 2014). Figure 13 depicts the smart grid investment problem, illustrating the business models and institutional settings that create (or enable to exploit) three municipal economic values usually not accounted for (i.e., RE connection co-ordination, inward investment stimulus, and municipal supplier DR management).

Risk and uncertainties related to smart grid deployment can take many forms, and affect multiple categories: markets, users, data and information, supply mix, policy, investment conditions, and networks (cf. Connor et al. 2018, p. 1).

Cambini et al. (2016) reviewed 459 innovative smart grid projects in 30 European countries realized between 2002 and 2014. Their focus was especially on (1) distribution sector concentration (i.e., the level of market concentration in the power distribution sector); (2) regulatory mechanisms (i.e., the capacity of the regulatory scheme to provide incentives to DSOs for increasing either their productivity or cost efficiency); (3) innovation–stimulus measures (i.e., the mechanisms designed by regulatory authorities to stimulate the implementation of innovative SG pilot projects).

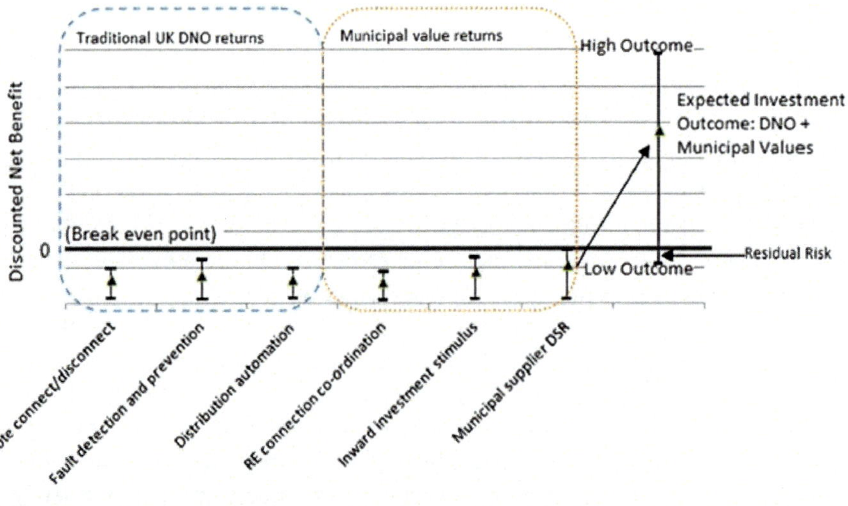

Fig. 13 The smart grid investment problem (Hall and Foxon 2014, p. 606)

The insights are useful for reflecting on the need for regulatory reform to stimulate innovation. These authors find that in countries with lower market concentration, DSOs invest on average much more (by a factor of 2) in SG than if the market concentration is high. Second, incentive-based regulatory schemes were found to also spur SG investment/innovation significantly more than cost-based regulation (by a factor of about 1.5). Finally, the adoption of innovation–stimulus mechanisms by regulation (e.g., adoption of an extra WACC or adjusted revenues) is rather successful in promoting SG investments (difference of a factor of 2.5).

DSO/DER regulation can be effected in different ways (Agrell et al. 2013):

1. *Integrated regulation*. An integrated DSO/DER entity considers the total cost of undertaking smart grid investments; the regulator, who does not know the costs but only forms expectations, makes a take-it-or-leave-it offer to the integrated entity.
2. *Decentralized regulation*. In this variant, the regulator makes an offer to the DSO. The latter decides what to offer to the DER investor. If the DSO investment costs exceed the budget alone, no investment can take place. The regulator's desire to limit the DSO–DER information rents will generally lead to underinvestment. Moreover, the DER is rationed more harshly than the DSO, because the DSO must pass on information rents to the DER.
3. *Centralized independent regulation*. The regulator signs separate and independent contracts with the DSO and DER, and rations to lower information rents. It implies a unilateral commitment by the regulator to finance the investment irrespective of the coordination in the supply chain.
4. *Centralized conditional regulation*. In this variant, the regulator offers the investment possibility to the DER investor first; if that one accepts and undertakes the investment, she also offers the investment to the DSO. An obvious advantage of this variant is that a situation where the DSO invests but the DER investor does not is avoided. Overall, the outcome of this arrangement may often be that the regulator should refrain from any investment to begin with (i.e., much like in the variant with individual regulation). More specifically, it is assumed that the regulator can make a conditional regulation in the sense that a separate contract is offered to the DSO and the DER investor first; an accepted contract by one party becomes only valid if the other party accepts the contract it was offered as well, thus avoiding the losses inherent in the unconditional centralized variant that arise when only one party accepts.

The interplay between smart meters, smart grids, and demand response can be analyzed, e.g., for the case of automated load control. The provision of real-time information about grid usage and nodal prices in distribution grids without any load-side application is useless, whereas installing DR without smart metering and information provision devices will not enhance social welfare (Agrell et al. 2013, p. 662).

Figure 14 depicts a model to systematically analyze DER investment scenarios. *Case A* depicts the integrated case where the unbundling requirement on the DSO is relaxed. This allows the DSO to undertake direct investments in DER activities. *Case B* (separate decentralized incentives) depicts a decentralized DSO where the regulator contracts with the DSO and delegates the coordination of DER investments to the DSO. In *Case C* (separate and centralized incentives), the regulator coordinates the DER investments directly via centralized incentives.

Both DSOs and owners of DER can undertake the investments considered useful and profitable (technologies, measurement equipment, protective devices, etc.), and social welfare gains may materialize either when both parties invest (*complementarity*) or when it is sufficient if one of them invests (*substitutability*).

In a smart grid system, there are totally new relations between the economic actors—not just suppliers and consumers of energy (services) but also prosumers (producer–consumers (Fig. 15)). Overall, this transition will also create major shifts in consumer behavior, lifestyles, and culture (Bigerna et al. 2016).

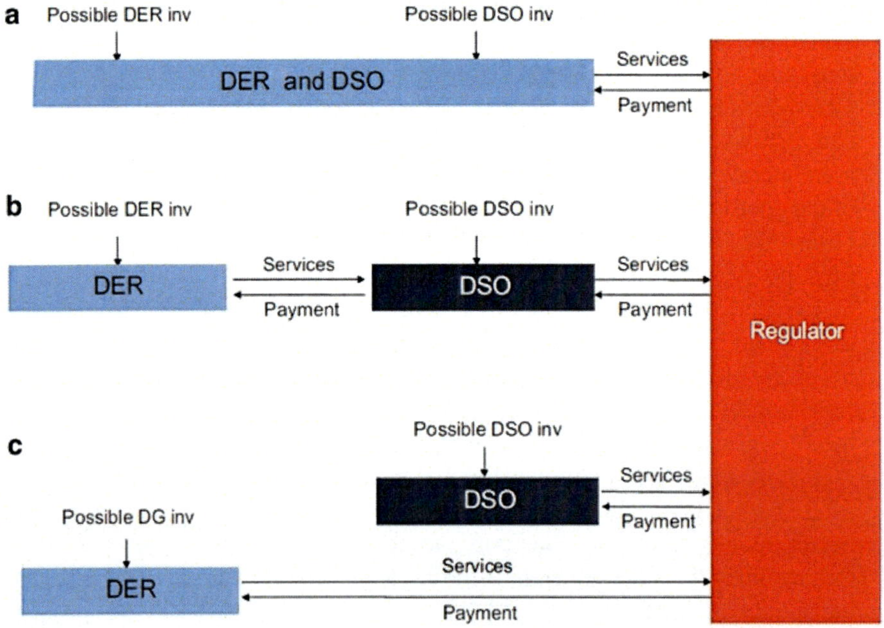

Fig. 14 DER and DSO investments and regulatory organization/delegation (Agrell et al. 2013, p. 661)

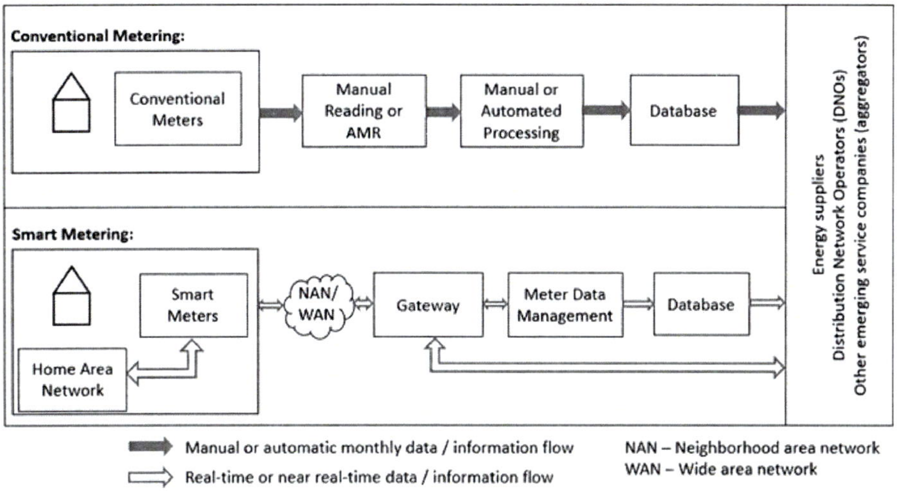

Fig. 15 Conventional versus smart metering, unilateral versus bilateral information flows and actors involved (Ekanayake et al. 2012)

6 Different Time Scales of Smart Grid Optimization and Impacts

6.1 Short Term Versus Long Term

The time domain is particularly important, as smart grid technologies—combined with new management and optimization techniques—not only enable much higher resolution analyses and problem-solving but also impose higher requirements on energy system modeling (see also the chapter "Modeling Smart Grid Systems"), ICT infrastructure, and computational power.

In the very short term, smart solutions to demand response and dispatching—and the increased flexibility and smartness in the system—enable an unprecedented matching of supply and demand.

In the longer term, it remains to be seen whether that leads to lower risk for investment decision-making, which is likely also much affected by radical innovation and poor or lagged regulation, among other influencing factors on stability and resiliency of the system.

Well-designed electricity markets and regulatory schemes can help to enhance the flexibility of the power system, and provide a level playing field for flexibility providers offering services (e.g., DR, generation and storage capacity, energy, reactive power) that enable the efficient and smooth operation and evolutionary further development of the smart grid system. The design and implementation of 'faster markets' (i.e. such with higher temporal resolution and response times) avoids unnecessarily high pricing (e.g. triggered by less flexible resources). It also enhances the possibilities for the integration or higher shares of VRES (Cruz et al. 2018, p. 348).

6.2 *Machine Learning and Energy Economics*

Artificial intelligence (AI), and especially its subdomain ML, is about to become a key enabler of a more complex and considerably more data-driven energy industry. The industry is changing rapidly and dramatically in the way energy supply, trading, and consumption are organized and managed. AI can outperform traditional approaches and models in terms of controllability (autonomously by smart software), handling of big data, smart grid system operation/management, predictive maintenance, cybersecurity, energy efficiency optimization, and much more (Cruz et al., Ahmad et al. 2021).

Smart grids offer new fields of application for ML techniques also with regard to the economic analysis of actors' behaviors and present and future behaviors of energy markets. Examples include demand forecasting, trading strategies for aggregated entities, analysis of (e.g., macro or energy) trends, energy price forecasting, and risk management (Ghoddusi et al. 2019). While economists are often concerned about the limitations of ML in terms of lacking theoretical foundation and causality inference, others are excited about the enormous potential for new ML applications in the energy domain, and the opportunities and benefits offered.

The most popular approaches used so far in the energy economics literature seem to be vector machines (SVMs), artificial neural networks (ANNs), and genetic algorithms (GAs).

A major difference between ML and traditional econometric modeling and analysis is the capability (and superiority) of ML algorithms in handling and analyzing large amounts of both structured and unstructured data, and to enable fast decisions or predictions. ML models neither require any structural modeling—i.e., assumptions of functional forms describing the interaction between variables—nor about stochastic distributions. ML has been shown to dominate statistical approaches in terms of learning more complex feature representations, so that, e.g., neural networks, support vector regression (SVR) and random forests typically outperform linear regression, lasso regression, or Box–Cox transformation regression, at least in terms of electric load forecasting. Still, there are also risks involved in using ML due to imperfect knowledge of the data or poor understanding of the algorithm/s used (many of which are 'black box' algorithms; cf. Kell et al. 2021).

In smart grids, where new data can in principle be provided in real time, online learning methods can be used to raise the accuracy of predictions. Online learning is particularly useful in the case of non-stationary data, and also for time series data where the recalculation of the algorithm would take a prohibitively long time. In contrast, offline learning methods need to be retrained each time new data has become available, leading to interruptions in the modeling.

Among the most frequent applications of ML in the energy economics literature—predicting energy prices (43%), predicting/modeling energy consumption/demand (39%), model calibration (2%), trading strategies (2%), structure of energy systems (7%), policy analysis (6%), and data management (1%) (cf. Ghoddusi et al. 2019, p. 712)—it can be expected that there will be some shifts in relative shares (and some new topics being added) when SG increasingly becomes a reality. Also, combining ML with econometric models, which has been a common approach aside from combining ML approaches (so-called 'ensemble approaches'), many new fields of application can be thought of.

In the prosumer analysis domain, which rises in importance as the number and relevance of prosumers will increase, ML can be used in predicting the energy demand and for identifying consumption patterns in prosumer households. Moreover, it can be used in solving the energy trading problem among prosumers (peer-to-peer trading, P2P) of a future electricity distribution system. Wang et al. (2021), for instance, propose a fully distributed energy trading framework based on ML to optimize the load and price prediction accuracy and energy trading efficiency.

Chapter 7 of this textbook is dedicated to smart grid business analytics, distinguishing among *descriptive* (business intelligence tools and processes for ex-post analysis), *predictive* (revealing patterns in the data useful for the prediction of future events), and *prescriptive* analytics (comprising the former ones and providing optimization-based guidance for decision-makers).

7 Case Study: The UK Smart Grid Initiative

In the United Kingdom, substantial progress has been made over the last years in the deployment of smart grids (Jenkins et al. 2015). A special focus has been put on the distribution grids, where the perception is that early action is needed. Among the expected benefits are not only the reduction in grid losses but also to enable distribution grid operators to better manage their carbon footprint.

Among the main actors involved are the government (Department of Energy and Climate Change, DECC), the regulatory authority (Ofgem), various grid operators, equipment manufacturers, and academia. Considerable investment has been made in smart grid research, development, and demonstration (RD&D) projects. Projects have been realized through several initiatives, such as the Office of Gas and Electricity Markets price control model (of Ofgem), which puts a lot of emphasis on supporting grid innovation. Moreover, the (competitive) funds Low Carbon Networks Fund (LCNF) and its successor fund, the Electricity Network Innovation Competition (ENIC) fund, provide the funding for grid operators to realize innovation projects and to test new smart grid technologies and solutions (Jenkins et al. 2015, p. 413). Smaller funds that also support to some extent smart grid projects are the Innovation Funding Incentive (IFI) and its successor, the Network Innovation Allowance (NIA).

The UK has also started to roll out smart meters, with the aim to improve the grid management possibilities and to enable smarter energy demand response and more customer engagement than without, thus harnessing the demand side for the enhanced control and optimization of the power system.

Energy demand has been, and will be, much affected by the market diffusion of electric heat pumps, electric vehicles, private household and community microgeneration, penetration of small-scale wind power generation, and other developments, all having an impact on the load on the distribution networks. DSOs need to be empowered to be able to actively manage intermittent and 2-way power flows. The list of emerging new requirements for the grid/DSO is a long one, and timely action is paramount.

Some future directions envisaged for the development of the smart grid in Great Britain include the demonstration of medium voltage direct current technology to bring more controllable connections between sections of the power grid, which would raise the amount of renewable power generation that can be connected to the grid. It further comprises wider applications of advanced ICT in transmission grids, which involves fiber optic cables instead of copper hard wiring, which will enhance controllability, reduce environmental impact, and increase station safety.

Yet another element in the strategy is the national rollout of smart meters (for both electricity and gas) in all homes and most small businesses, involving the replacement of some 47 billion meters and costs of some £8.6 bn. Supplier benefits were estimated at more than £6 bn, of which avoided meter reading accounts for some £2.6 bn and reduced inquiries and customer overheads £1.13 bn (Jenkins et al. 2015). Consumer benefits were estimated at some £6.4 bn, composed of mainly energy cost savings (£4.2 bn) and load-shifting/TOU tariffs (£1.1 bn) (cost savings partly occurring upstream but or assumed to be passed on downstream to the consumers). For grid operators, smart metering is expected to provide benefits in terms of network planning, network operation, and demand management. *Network planners* can benefit from the analysis of load and voltage profiles obtained from smart meters (instead of standard/synthetic load profiles), better asset utilization on the distribution grid level (enabling, e.g., deferral of asset replacement), and providing a more accurate basis for predictions of future voltage and demand operating ranges in light of the expected rapid diffusion of heat pumps and electric vehicles. *Network operators* can benefit from better local outage detection and shorter system restoration times, also enabling them to communicate reinstatement of disconnected supplies using smart meter communication channels (greatly reducing reliance on customer calls both for local fault or outage detection). Finally, *demand management* will benefit greatly from smart meter installations in terms of enhancing demand response, e.g., by communicating dynamic pricing levels, or (incentive-driven) direct load control (see also the chapter "Demand Side Management").

Obviously, such estimates of costs and benefits are subject to high uncertainty. Connor et al. (2018) systematically identified risks and uncertainty related to the UK smart grid deployment based on a detailed expert stakeholder analysis. They found that many of the risks identified arise from the increasing complexity of the system and of potential solutions that come along the evolutionary process of steadily increasing smartness. A major source of risk identified was continued inconsistency in the aims and objectives of different regulatory bodies (esp. Ofgem and DECC, now

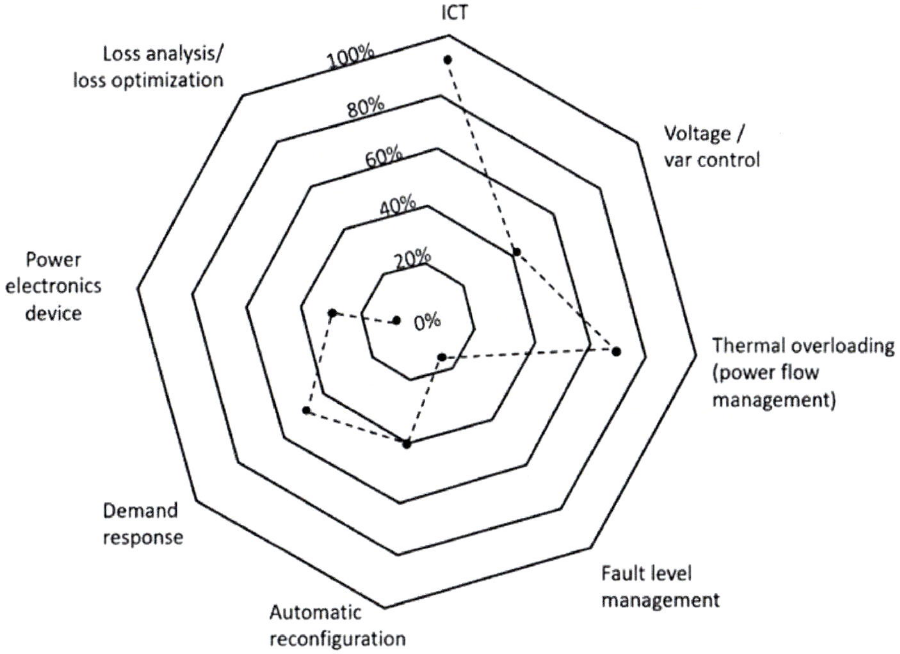

Fig. 16 Technical foci of the tier-2 smart grid projects realized in the UK (Jenkins et al. 2015, p. 418)

BEIS). As a remedy, there is a need for a virtuous co-evolution of the regulation (and the regulator) along with that of technology and market services (and thus also societal needs and lifestyles). Another important risk perceived as difficult to mitigate is that of 'broken value chains' (i.e., problems in 'value-stacking', i.e., if new technologies and services cannot be monetized because the value created is dispersed and difficult to aggregate and/or share cost-effectively among the stakeholders). Additional network costs, and the impact on consumer bills if passed on to them, were also seen as a significant risk. Finally, creating value from customer/consumer data was seen as a risk for acceptance of SG solutions on the consumer side. Overall, however, all these risks and uncertainties identified were not used as an argument to continue with the current solutions, as these will become less effective and more expensive as the sustainable energy transition progresses.

Figure 16 shows the shares of projects by technology among all tier-2 projects. It shows that ICT technologies played the most important role, as more and more sensors and remote terminal units are installed in the distribution grid for improving visibility and to enable grid control and automation. More than half of the projects were focusing on the active management of the power flows in the low- and medium-voltage grids, exploring concerns on voltage and thermal constraints. Fault-level management has so far seen limited attention. Automatic network reconfiguration can be realized by means of automatic switches and power electronics, and is considered a very effective means for demand balancing and voltage stabilization in smart grids.

To date, the focus has been mainly on the distribution grids, including real-time information flows and interaction between suppliers and consumers based on improved ICT, active power flow management, and DSM. The national rollout of smart meters boosts the smart grid development, as it enables more efficient network planning and operations, and customer engagement through DR.

Review Question

- What are the main potentials of smart grids to enhance economic efficiency?
- Can you distinguish between the different flexibility options, their relative merits, and the competition between them?
- Do you have a clear idea of types of actors, DER and concepts involved in a SG system, and more broadly, a sense of the economic implications of the '4Ds'?
- Do you have a basic idea on local energy market design issues? Did you obtain a basic understanding of the investment needs under uncertainty?
- Did you develop a sense for the many new fields of application for machine learning techniques arising with the realization of smart grids?

References

Agrell PJ, Bogetoft P, Mikkers M (2013) Smart-grid investments, regulation and organization. Energy Policy 52:656–666

Ahmad T, Zhang D, Huang C, Zhang H, Dai N, Song Y, Chen H (2021) Artificial intelligence in sustainable energy industry: status quo, challenges and opportunities. J Clean Prod 289:125834

Bigerna S, Bollino CA, Micheli S (2016) Socio-economic acceptability for smart grid development—a comprehensive review. J Clean Prod 131:399–409

Cambini C, Meletiou A, Bompard E, Masera M (2016) Market and regulatory factors influencing smart-grid investment in Europe: evidence from pilot projects and implications for reform. Util Polic 40:36–47

Connor P, Axon C, Xenias D, Balta-Ozkan N (2018) Sources of risk and uncertainty in UK smart grid deployment: an expert stakeholder analysis. Energy 161:1–9

Council for European Energy Regulators (2019) CEER Consultation on Dynamic Regulation to Enable Digitalisation of the Energy System. Technical report, Brussels, Belgium. Conclusions Paper, Ref C19-DSG-09-03, 10 Oct 2019

Cruz MRM, Fitiwi FZ, Santos SF, Catalão JPS (2018) A comprehensive survey of flexibility options for supporting the low-carbon energy future. Renew Sustain Energy Rev 97:338–353

Crampes C, Trochet J-M (2019) Economics of stationary electricity storage with various charge and discharge durations. J Energy Storage 24:100746

De Oliveira-De Jesus PM, Henggeler Antunes C (2018) Econmic valuation of smart grid investments on electricity markets, Sustain Energy, Grids and Networks 16:70–90

Eid C, Codani P, Perez Y, Reneses J, Hakvoort R (2016) Jun). Managing electric flexibility from distributed energy resources: a review of incentives for market design. Renew Sustain Energy Rev 64:237–247

Ekanayake JB, Jenkins N, Liyanage K, Wu J, Yokoyama A (2012) Smart grid: technology and applications. Wiley, New York

Faerber LA, Balta-Ozkan N, Connor PM (2018) Innovative network pricing to support the transition to a smart grid in a low-carbon economy. Energy Policy 116:210–219

Fuhrmann J, Madlener R (2020) Evaluation of synergies in the context of European multi-business utilities. Energies 13(24):6676

Geidl M, Koeppel G, Favre-Perrod P, Klockl B, Andersson G, Frohlich K (2007) Energy hubs for the future. IEEE Power Energy Mag 5(1):24–30

Ghoddusi H, Creamer GG, Rafizadeh N (2019) Machine learning in energy economics and finance: a review. Energy Econ 81:709–727

Giulietti M, Grossi L, Trujillo Baute E, Waterson M (2018) Analyzing the potential economic value of energy storage. Energy J 39(SI1):101–122

Green R, Staffell I (2017) "Prosumage" and the British electricity market. Econ Energy Environ 6(1):33–49

Gui EM, Diesendorf M, MacGill I (2017) Distributed energy infrastructure paradigm: community microgrids in a new institutional economics context. Renew Sustain Energy Rev 72:1355–1365

Gui EM, MacGill I (2018) Typology of future clean energy communities: an exploratory structure, opportunities, and challenges. Energy Res Soc Sci 35:94–107. Energy and the Future

Gui EM, MacGill I (2019) Chapter 7—Consumer-centric service innovations in an era of self-selecting customers. In: Sioshansi F (ed) Consumer, Prosumer, Prosumager. Academic Press, London, San Diego, Cambridge, MA, Oxford, pp 127–151

Hall S, Foxon TJ (2014) Values in the smart grid: the co-evolving political economy of smart distribution. Energy Policy 74:600–609

Hanna R, Ghonima M, Kleissl J, Tynan G, Victor DG (2017) Evaluating business models for microgrids: interactions of technology and policy. Energy Policy 103:47–61

Hirth L, Ueckerdt F, Edenhofer O (2015) Integration costs revisited-an economic framework for wind and solar variability. Renew Energy 74:925–939

Jamasb T, Llorca M (2019) Energy system integration: economics of a new paradigm. Energy J 8(2):7–28

Jenkins N, Long C, Wu J (2015) An overview of the smart grid in Great Britain. Engineering *1*(4):413–421

Jesus PDO-D, Henggeler Antunes C (2018) Economic valuation of smart grid investments on electricity markets. Sustain Energy Grids Netw 16:70–90

Karami M, Madlener R (2021) Business model innovation for the energy market: joint value creation for electricity retailers and their customers. Energy Res Soc Sci 73:101878

Kell AJ, McGough AS, Forshaw M (2021) The impact of online machine-learning methods on long-term investment decisions and generator utilization in electricity markets. Sustain Comput: Inf Syst 30:100532

Lineweber DC (2011) Understanding residential customer support for—and opposition to—smart grid investments. Electr J 24(8):92–100

Lu X, Li K, Xu H, Wang F, Zhou Z, Zhang Y (2020) Fundamentals and business model for resource aggregator of demand response in electricity markets. Energy 204:117885

Lüth A, Zepter JM, Crespo del Granado P, Egging R (2018) Local electricity market designs for peer-to-peer trading: the role of battery flexibility. Appl Energy 229:1233–1243

Luthra S, Kumar S, Kharb R, Ansari MF, Shimmi S (2014) Adoption of smart grid technologies: an analysis of interactions among barriers. Renew Sustain Energy Re 33:554–565

Madlener R, Specht JM (2018) Business opportunities and the regulatory framework. In: R.E. Hester and R.M. Harrison (Eds.), Energy Storage Opt Their Environ Impact, Royal Society of Chemistry, Ch. 9, pp 296–320. (ISBN 978-1-78801-399-4)

MIT Energy Initiative (2013) The MIT Utility of the future study, White Paper, Massachusetts Institute of Technology, Boston, MA

Morstyn T, Farrell N, Darby SJ, McCulloch MD (2018) Using peer-to-peer energy-trading platforms to incentivize prosumers to form federated power plants. Nat News 3:94–101

Müller-Stewens G, Brauer M (2009) Corporate Strategy & Governance. Wege zur nachhaltigen Wertsteigerung im diversifizierten Unternehmen. Schäffer-Poeschl, Stuttgart, Germany
Nolting L, Schuller V, Gaumnitz F, Praktiknjo A (2019) Incentivizing timely investments in electrical grids: analysis of the amendment of the German distribution grid regulation. Energy Policy 132:754–763
Oberst C, Madlener R (2014) Prosumer preferences regarding the adoption of micro-generation technologies: empirical evidence for German homeowners. FCN Working Paper No 22/2014, Institute for Future Energy Consumer Needs and Behavior, RWTH Aachen University, Aachen, Germany, 2014, Dec (revised Sep 2015)
Oberst C, Schmitz H, Madlener R (2019) Are prosumer households that much different? Evidence from stated residential energy consumption in Germany. Ecol Econ 158:101–115
O'Malley M, Kroposki B, Hannegan B, Madsen H, Andersson M, D'haeseleer W, McGranaghan M, Dent C, Strbac G, Baskaran S, Rinker M (2016) Energy system integration: defining and describing the value proposition. Report no tp5d00-66616, Golden, CO
Pagani GA, Aiello M (2015) From the grid to the smart grid, topologically. Physica A: Stat Mech Appl 449:160–175
Parag Y, Sovacool B (2016) Electricity market design for the prosumer ERA. Nat Energy *1*, 16032
Rosen C, Madlener R (2013) An auction design for local reserve energy markets. Decis Support Syst 56:168–179
Sabadini F, Madlener R (2021) The economic potential of grid defection of energy prosumer households in Germany. Advances in Applied Energy, 4: 100075. https://doi.org/10.1016/j.adapen.2021.100075
Schill W-P, Zerrahn A, Kunz F (2017) Prosumage of solar electricity: PROS, CONS, and the system perspective. Econ Energy Environ Policy 6(1):7–31
Schmitz H, Madlener R (2021) Mechanisms for Rebound Effects of Solar Electricity Prosuming in Germany, FCN Working Paper No. 10/2021, Institute for Future Energy Consumer Needs and Behavior, RWTH Aachen University, Aachen, Germany
Shomali A, Pinkse J (2016) The consequences of smart grids for the business model of electricity firms. J Clean Prod *112*(5), 3830–3841
Sioshansi R, Denholm P, Jenkin T (2012) Market and policy barriers to deployment of energy storage. Econ Energy Environ Policy 1(2):47–63
Specht JM, Madlener R (2019) Energy supplier 2.0: a conceptual business model for energy suppliers aggregating flexible distributed assets and policy issues raised. Energy Policy 135:110911
Toft MB, Schuitema G, Thøgersen J (2014) The importance of framing for consumer acceptance of the smart grid: a comparative study of Denmark, Norway and Switzerland. Energy Res Soc Sci 3:113–123
Toffler A (1980) The Third Wave, Penguin Press
van Leeuwen G, AlSkaif T, Gibescu M, van Sark W (2020) An integrated blockchain-based energy management platform with bilateral trading for microgrid communities. Appl Energy 263:114613
Vergados DJ, Mamounakis I, Makris P, Varvarigos E (2016) Prosumer clustering into virtual microgrids for cost reduction in renewable energy trading markets. Sustain Energy Grids Netw 7:90–103
Verkade N, Höffken J (2018) The design and development of domestic smart grid interventions: insights from the Netherlands. J Clean Prod 202:799–805
Wang N, Li J, Ho S-S, Qiu C (2021) Distributed machine learning for energy trading in electric distribution system of the future. Electr J 34(1):106883. Special Issue: Machine Learning Applications To Power System Planning And Operation
Zame KK, Brehm CA, Nitica AT, Richard CL, GDS III (2018) Smart grid and energy storage: Policy recommendations. Renew Sustain Energy Rev *82*(1), 1646–1654
Zepter JM, Lüth A, Crespo del Granado P, Egging R (2019) Prosumer integration in wholesale electricity markets: synergies of peer-to-peer trade and residential storage. Energy Build 184:163–176
Zweifel P, Praktiknjo A, Erdmann G (2017) Energy economics: theory and applications, 1st edn. Springer, Berlin, Heidelberg, New York

Demand Side Management

Joan Batalla-Bejerano, Elisa Trujillo-Baute, and Reinhard Madlener

Learning Objectives

- Be able to describe the main motivation and technical options for demand side management (DSM)
- Be able to understand the potentials and future perspectives of DSM
- Be able to classify and describe price- and incentive-based demand response (DR) programs
- Be able to understand demand flexibility markets
- Be able to describe the emerging new business actors, their aims and roles, and (new) opportunities related to DSM

J. Batalla-Bejerano
Funseam and Chair of Energy Sustainability IEB-UB, Funseam, C/ John Maynard Keynes, 1-11, 08034 Barcelona, Spain
e-mail: jbatalla@funseam.com

E. Trujillo-Baute (✉)
University of Lleida and Chair of Energy Sustainability IEB-UB, Funseam, C/ John Maynard Keynes, 1-11, 08034 Barcelona, Spain
e-mail: elisa.trujillo@udl.cat

R. Madlener
Chair of Energy Economics and Management, School of Business and Economics / E.ON Energy Research Center, Institute for Future Energy Consumer Needs and Behavior (FCN), RWTH Aachen University, Aachen, Germany
e-mail: RMadlener@eonerc.rwth-aachen.de

C. Dinther et al. (eds.), *Smart Grid Economics and Management*,
Lecture Notes in Energy 51, https://doi.org/10.1007/978-3-030-84286-4_3

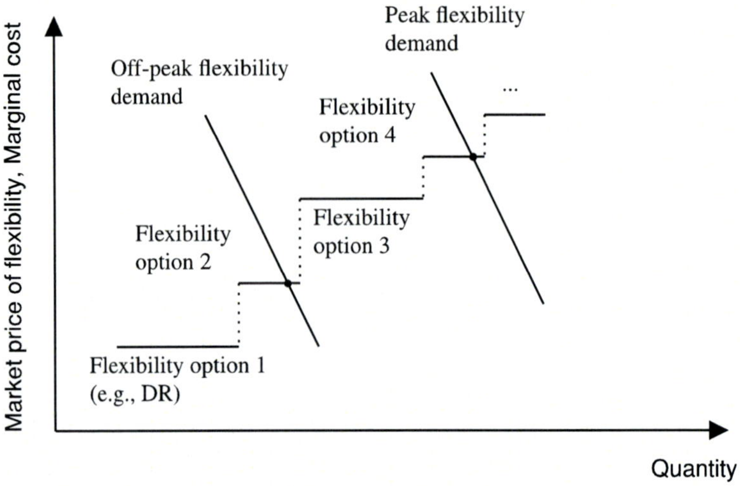

Fig. 1 Merit order of flexibility options in a smart grid (viewed as a market for flexibility with supply and demand)

1 Economic Considerations

To respond to the variability of supply that characterizes power generation from renewable energy sources (RES) a range of potentially competing options exists, including DR, grid expansion, energy storage, enhanced flexibility of power generation, and improved operational practices. For a better understanding, the set of alternatives that can provide responsive energy over various timescales can be grouped into supply side, demand side, and grid-related ones (Denholm and et al. 2010; IRENA 2018). While supply-side flexibility is closely related to the performance of the technologies comprising the generation units of a power system, demand-side flexibility refers to specific types of demand-side management where the demand pattern could be shifted to better match electricity supply (IRENA 2018).

Although the objective of this chapter is far from an in-depth explanation of the technical characteristics of each option, it is relevant to point out that each alternative option to enhance the flexibility of the energy system comes at different potentials, spatial and temporal availability, and costs, and thus fits better (or worse) for the different timescales and challenges related to the integration of growing shares of variable renewables. Figure 1 illustrates the "merit order of flexibility" in a stylized manner. Obviously, the costs of some of these options can be expected to come down over time due to learning and economies of scale effects.

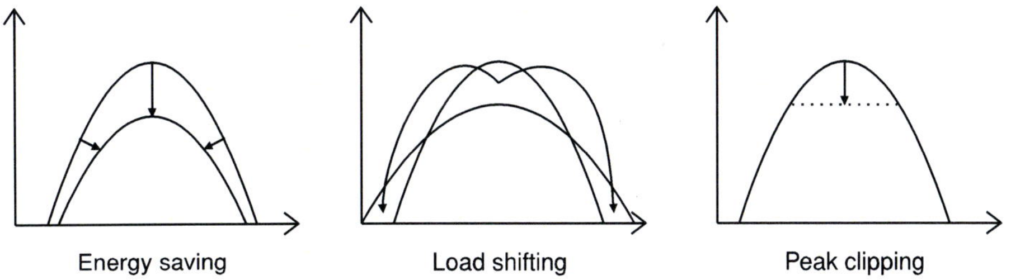

Fig. 2 Demand response and form of load curve obtained based on (Batalla-Bejerano et al. 2020, p. 4)

2 Demand Response Potential and Future Perspectives

With the aim of improving efficiency, reliability, and safety of the power system, intelligent technologies are being incorporated across the entire system, from power generation, transmission, and distribution networks, to electricity consumption at the premises of consumers. Within the context of the energy transition, one of the key objectives of smart technologies is an energy-efficient power grid to better manage volatile renewables and demand. DSM includes all the activities that target the modification of the consumer demand profile, in time or in shape, to make it better match the supply (Alizadeh et al. 2012).

Through different mechanisms, demand response allows consumers to play a role in the operation of the grid by adjusting their electricity consumption subject to price signals or long-term direct-control agreements with different objectives. Demand response programs can be introduced in the following ways (Fig. 2):

- **Peak Clipping.** To reduce consumption at times of peak demand. The hourly demand for electricity fluctuates throughout the day, with peaks normally around noon and in the early evening. During these periods, it is necessary to employ generation technologies with higher costs in order to meet demand. Because of this, one of the objectives of any energy policy is to reduce the relevance of these peaks in demand given their implications in economic and environmental terms and where the introduction of smart meters with their associated new functionalities offer the possibility of managing electricity demand, allowing greater efficiency in the markets in the years to come.
- **Load Shifting.** To move electricity consumption from one time period to another—normally with less expensive generation costs, such as advancing or deferring the use of an appliance to another time. The idea behind this strategy is that, by shifting the load to another period, the returns generated through energy cost savings or DR participation are greater than the potential well-being loss from the behavioral change. Unlike other energy cost-saving strategies, the issue tackled by the load shifting is the when rather than the how much. Hence, with load shifting, the overall energy consumption is the same but the moment when it is consumed changes.

- **Strategic Energy Saving.** To reduce energy consumption at all times. All efforts that imply reductions in the load curve entail multiple benefits in terms of competitiveness, security of supply—especially for those countries or regions strongly dependent on external energy resources—and environmental sustainability. Measures such as improved energy efficiency in buildings or in devices that consume energy at home (i.e., heating, cooling or appliances) are included in this area. As for the DR for the residential sector, the conservation strategy is to encourage households to modify their patterns of behavior in response to an indication of consumption and/or prices, leading to a reduction in consumption. Unlike peak clipping and load shifting, consumption reduction is not necessarily put into effect during a specific intraday period, such as times of peak demand, nor is consumption shifted to other periods.

Among these technical options, DR can be used in combination with energy storage to further reduce curtailment of variable renewable generation and independently of the strategy, the aim behind all of the options is to benefit from the huge potential that is inherent in DR programs.

According to the European Commission (2016), the theoretical potential of demand response, understood as a change in electricity consumption patterns in response to a signal, in 2016 adds up to about 100 GW and is expected to reach 160 GW in 2030. The highest share of DR potential lies in the residential sector that accounts for 23% of total energy consumption worldwide, placing it third after industry at 37% and transportation at 28% (IEA 2019).

However, some barriers need to be overcome before we can expect to see the wide-scale uptake of demand response solutions. Regulatory and market barriers seem to be the main obstacles. Regulatory support is key to the success of DR, given that if regulators do not succeed in structuring the market so that energy savings benefit the consumers in terms of final cost reductions, consumers have no compelling reason to implement DR programs. At the same time, removal of market barriers to demand side participation on the wholesale markets is crucial and the figure of the aggregator and its future role could help in the provision of new services and products that foster demand response.

Another challenge to be overcome is gaining consumer trust and encouraging consumer participation in this kind of DR initiatives. While the business sector is largely driven by economic reasoning, households are not necessarily so. Factors such as education, social norms, age, and culture are prevalent (e.g., Darby 2010; Hargreaves et al. 2010; Vine et al. 2013). Nevertheless, in a context where energy prices are continuously rising, the public's engagement with demand response will increase, as consumers are attracted by the financial and efficiency benefits that DR solutions can offer.

Consumer empowerment to participate and well-functioning flexibility markets are the two key components needed for a successful DSM. The first part of this section is devoted to exploring the measures to empower electricity consumer participation. The DR emerges as one of the main activities within the DSM, characterized by including changes in the consumption pattern in response to certain signals. Energy storage systems and RES self-production are also considered to be relevant

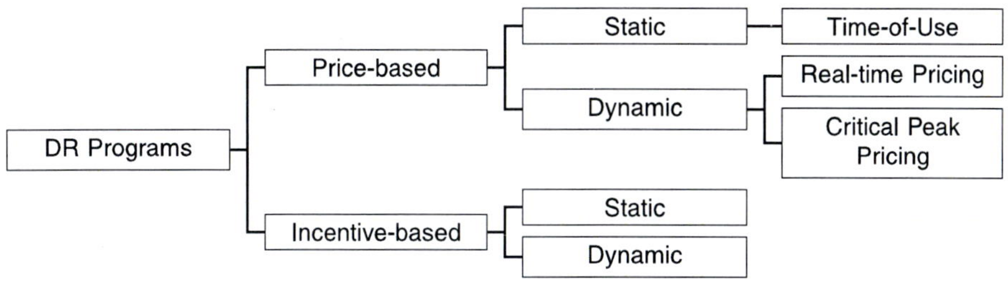

Fig. 3 Demand response programs

elements facilitating electricity consumer participation. The installation and use of interconnected home appliances are essential to the successful functioning of DSM, alongside with the production of the equipment needed and the setting of specific technical requirements to be fulfilled. An additional element influencing consumer empowerment is the energy service company (ESCO), acting as an intermediary in charge of facilitating the finance of energy-saving programs. Finally, given its prominent role in creating the necessary space for supply and demand interaction to achieve efficiency gains, in the second part of the section, a detailed description of flexibility market designs and functioning is provided, including both ancillary and balancing markets.

2.1 *Consumer Participation in Demand Side Management*

Demand response is usually defined as a change in the consumption patterns of electricity consumers in response to time-varying tariff structures or incentive payments in order to operate the electricity system in a more efficient and reliable manner (see also Chap. 2 on the elasticity of demand in light of enhanced flexibility). Hence, the principal DR activities include tariffs or programs designed to motivate the electricity use by end-user consumers, in response to changes in the price of electricity over time, or to give incentive payments designed to induce lower electricity use at times of high electricity demand or when facing reliability issues.

Depending on the offered motivation to shape consumer behavior, DR programs are typically classified in two main categories (Albadi and El-Saadany 2007): price-based programs and incentive-based programs (Fig. 3). While the price-based programs offer customers time-varying prices that are defined based on the cost of electricity in different time periods (Aghaei and Alizadeh 2013), incentive-based ones consist of programs that offer fixed or time-varying incentives (payments) to customers that reduce their electricity usage during periods of system stress (Mohagheghi et al. 2010).

With the aim of shifting consumption from periods where costs are high to periods where costs are low, in the price-based programs, the actual cost of electricity generation, transport, and supply is reflected in time-varying tariff structures. The price-based programs can be classified into static and dynamic time-varying pricing. The basic difference between these two types of price-based programs is that in the static pricing scheme, time-varying retail prices are present for pre-determined hours and days, while dynamic prices are allowed to change on short notice.

Within the price-based programs, the static time-varying prices (usually referred to as time of use (ToU)) is the application of flat pricing to different time periods. Under the ToU pricing scheme, prices are retained fixed within different pricing periods, which can be different hours during a day or different days during the week. For instance, within the day, ToU typically include two or three different price increasing periods: off-peak, (mid-peak), and peak. The dynamic pricing programs are typically represented in two basic types: real-time pricing (RTP) and critical peak pricing (CPP). While in RTP, the price changes every hour following exactly the changing market conditions, the CPP allows for the retailer to occasionally declare an unusually high retail price for a limited number of hours and to charge those prices to consumers. Dynamic pricing can be more complex than static pricing, but it has been shown that it can capture a far larger share of the variation in wholesale prices (Borenstein et al. 2002).

In the incentive-based programs, consumers obtain (fixed or time-varying) payments for participating and for reducing their load when required. Consumer enrollment and response are voluntary, although depending on the terms of the contracts a financial punishment might exist in the case of failing (no participation) when events are declared. The incentive-based DR programs can be categorized into classical, with consumers receiving a participation payment (usually in the form of bill credits or discount rates), and market-based, where participants are rewarded with money for their performance in terms of the amount of electricity reduced during critical conditions.

The most widely used incentive-based programs are the interruptible Load (IL) schemes, where pre-defined incentives are provided to participating consumers. An IL program considers curtailment options like, for instance, curtailing a specific part of the electric load or curtail the total consumption to reach a predetermined level. While partial load interruption is more often used in the context of domestic consumption, the total curtailment is mostly used in the case of industrial consumers. In addition, these DR programs can provide a rate of discount or bill credit in exchange to reduce the load during system's emergency, and consumers that do not respond to these options receive a number of penalties as defined in the agreed terms of conditions (Albadi and El-Saadany 2008; Faria and Vale 2011). Salah et al. (2017) provide an alternative angle on the subject matter by developing a morphological taxonomy of different rate design features which jointly span the design space for all different energy pricing schemes.

The savings on the energy bill achievable by consumers through the participation in DR programs during peak hours are among the most often emphasized benefits of these programs. And from the consumer point of view, these benefits constitute the main element stimulating consumer participation in DSM programs. In addition, because expensive peak plants have to be ramped up with a lower frequency, as a result of DR a market-wide price reduction can be expected. And with respect to grids, with DR investment in infrastructure can be deferred and the management of grid congestion issues can be dealt with a number of different tools.

RES self-generation is considered an important facilitator of electricity consumer participation in the context of DSM. Consumer awareness and choice relative to energy management and self-generation options is growing. Consumers have multiple choices at their fingertips to explore, including rooftop solar PV, smart appliances, electric vehicle (EV) charging options, heating, ventilation, and air conditioning (HVAC), water heating, and solar PV plus storage. All aspects of homes connected to the grid can accommodate customer preferences and comfort. Because these behind-the-meter elements can be read in near real-time, they can be used to generate lower or more predictable energy bills for consumers. Usage insights can also be used to help drive additional products and services to improve the customer experience and ultimately incentivize their participation in DSM programs.

Another relevant element facilitating electricity consumer participation in DSM are energy storage systems. Multiple storage technologies can be used in both households and industrial facilities to alter the electricity load from times with surplus energy to be released for later use, i.e., load shifting. Particularly in the context of RES self-production, it has been demonstrated that the combination of small-scale storage with DSM significantly improves the local use of photovoltaics (PV), thus increasing the PV value for the user. It is foreseen that this combination will play an important role in the future smart grids (Castillo-Cagigal et al. 2010). Considering distributed energy resources (DER), by storing electricity when local production exceeds consumption, future needs can be satisfied with electricity coming from this local source. From the grid point of view, this corresponds to anticipating a part of the consumption, for instance during midday sunny hours instead of during the evening load peak. As a result, it can create value for the grid, part of it being shared with end-users by reducing their consumption costs without modifying their electricity demand patterns.

2.2 The Role of Smart Grids and Smart Devices

Currently, DSM is increasingly viable by means of highly efficient electrical appliances that can be remotely controlled (Hinnells 2008). Many of these devices can also benefit the grid when collectively managed, by minimizing, shifting or optimizing energy use within the home. In addition to premise-level optimization technologies, distributed energy resource management systems are being piloted and deployed by suppliers and third-party aggregators to control fleets of customer assets that meet a variety of grid needs. It is apparent that the installation and use of interconnected home appliances are essential to the successful functioning of DSM, this under the pre-existent condition of having some guarantee that the production of the equipment (sensors, smart meters, and control systems, among others) and the setting of specific technical requirements are fulfilled.

Combined with cost-reflective price signals, such as dynamic pricing, DSM systems can provide a way for consumer empowerment and facilitate better coordination of energy usage in ways that reduce the cost of delivering electricity, which supports a more affordable, reliable, and sustainable grid over time. Thanks to smart device technology, consumers can now understand how consumption patterns impact their energy bill, comfort level, and supply security. Consumers can see how energy usage during peak times is more expensive and shift their use during less expensive times of the day. In addition, data from smart meters allows suppliers to observe consumption patterns and load shapes, which can be used to improve blackout management and the overall flexibility of the grid.

2.3 The Role of Energy Service Companies

An additional element influencing consumer empowerment are ESCOs, acting as intermediaries in charge of facilitating the finance of energy saving programs. The leverage that comes from the evolution of smart grids is described in more detail below, in particular, the multiple roles of conventional ESCOs are summarized. These financing arrangements differentiate ESCOs from the traditional energy consultants or equipment suppliers. ESCOs have the know-how required to provide key services and solutions for achieving significant cost reductions while dealing with various market-related barriers remaining. Some of the most common functions of ESCOs include handling projects, managing or mobilizing financial and non-financial resources, as well as undertaking installation and maintenance work.

ESCOs typically carry out their economic activity under two main contracting mechanisms: energy performance contracting (EPC) and energy supply contracting (ESC). By linking their benefit to the performance of their implemented projects, ESCOs assume performance risks under EPC. With this risk acceptance, ESCOs ultimately incentivize themselves to deliver savings-oriented solutions for consumers and for society, with the consequent overall welfare gains. Under the ESC concept, the ESCO is only remunerated for the useful energy output, that is, it supplies useful energy (such as electricity, heat, or steam) under a long-term contract to a building owner or building user. It is, therefore, in the interest of the ESCO to reduce the final energy demand, aligning once again the incentives of both, energy consumer and service provider.

3 Business Opportunities and New Market Actors

In the context of demand side management, new services emerged and traditional actors, like retailers, are becoming more active while facing challenges from new actors competing for customer service provision, namely, the aggregators and the ESCOs. In this section, we explain the roles and the interplay of these actors, along with the roles equipment and appliance manufacturers have in providing the required elements for advanced services.

Power aggregation can significantly speed up the integration of renewable power generation from intermittent sources, improve demand flexibility, and decrease the dependence on renewable energy support policies. However, single individual, commercial or domestic consumers would only have a limited impact towards the achievement of more aggregation and market integration. The use of distributed generation, demand side response, and different forms of storage can only be effective through the coordinated actions of massive amounts of diverse consumers and producers interacting in the markets.

Aggregators

Aggregators are formally defined as legal entities with the aim of optimizing, technically or economically, the production and consumption of an electricity system. The aggregation can involve generators and consumers, operating in one or multiple electricity markets. In other words, they are facilitators between the supply and demand sides of electricity markets, and under a design including the appropriate incentives, the wholesale electricity markets might benefit from aggregation. On the downstream side of the market, aggregators act on behalf of consumers through the use of technological solutions and information and communication technology (ICT) for optimization. In this business line, aggregators provide energy services for industrial, commercial or domestic consumers who own generation and storage units or can offer demand response. On the upstream side of the market, energy aggregators offer value to multiple market players to optimize their portfolios and for balancing and congestion management. In this business stream, the aggregator's services are provided to balance responsible partys (BRPs), distribution system operators (DSOs), transmission system operators (TSOs), and energy suppliers.

ESCO

One last option for the aggregator is to offer value to some downstream and upstream entity—this is the case of ESCOs. Demand response programs are opening up new opportunities for ESCOs to play a significant role in the operation and optimization of energy grids. After all, ESCOs have a unique competitive advantage: they can combine their energy management expertise with granular customer insights to help reduce or shift their clients' energy usage during peak periods in response to time-based tariffs or other incentives.

A context with an increasing demand side management will also offer a great way for ESCOs to extend energy services beyond raising energy efficiency—by contributing to grid stability, safety, and environmental sustainability—while reaping additional cost savings and opening up new revenue streams for their customers. This does not mean that ESCOs will necessarily enter into competition with other new market players, e.g., "aggregators". ESCOs can take advantage of upcoming energy monitoring platforms which enable their clients to participate in demand response programs without having to take an aggregator role. Consequently, an ESCO will only have to install the necessary equipment for demand response as it usually does for other purposes such as sub-metering.

Energy Supplier 2.0

The "Energy Supplier 2.0", already mentioned in Chap. 2, is a conceptual business model for energy suppliers aggregating flexible distributed assets introduced by Specht and Madlener (2019). The focus of new business models in the electricity supply market (and actually also other end-user energy supply markets) has to focus much more on the customer needs and potentials. The conceptual business model of the Energy Supplier 2.0 (that is based on the Business Model Canvas approach by Osterwalder and Pigneur (2010, cf. Chap. 8) sketches out a dedicated aggregator of flexible capacities on the household level. It reveals how a specific new energy business model can tap the potential of distributed flexible energy assets (including DR), and is thus also very relevant for new demand side management services offered by an ES2.0.

Virtual power plant operators

Virtual power plants (VPPs), as an orchestration of a collection of DER, have been proposed to reduce upstream power generation and transmission capacity needs, but also as a new business model enabled by smart grid technology. VPPs may include demand response aggregators and/or such aggregating battery capacity, both focusing on flexible loads. The concept where a VPP is formed through peer-to-peer transactions between self-organizing prosumers has been referred to as "federated power plant" (Morstyn et al. 2018). A VPP is a cloud-based distributed power plant that aggregates the capacities of heterogeneous DER for the purposes of enhancing power generation, as well as trading or selling power on the electricity market. Under a VPP, all the elements of the grid (sources, storage, building, district, transformers, etc.) are integrated and all communicate together, using standard TSO/DSO communication protocol, to choose the best management solution to provide energy services.

Prosumers

Prosumers can be autonomous from the grid or not. Those households that can make themselves independent from the grid throughout the year need to install sufficient renewable energy capacity and energy storage, as well as some smart home/home automation technologies for energy management including DR. Those prosumers who remain connected to the grid may be passive or active agents from the grid perspective, i.e., either providing some energy and/or ancillary services to the grid or not. Efficient demand management is paramount for maximizing the level of autarky that can be achieved at a reasonable cost.

Citizen energy communities

On the meso level between an individual household and the entire supply region of an energy provider, there are local, more or less self-sufficient, energy communities. With a multitude of potential legal arrangements (e.g., cooperative societies, public limited companies, or some small-medium size enterprise), energy communities integrates, aggregates, and coordinates distributed energy services for users and providers at the neighborhood level, facilitating the active involvement of citizens in the configuration and operation of the energy system, and its flexibility. Besides, energy communities can team up with other market players and jointly invest in energy assets, enabling the interactions in such a way that the full potential from DR is gathered for the shared benefit of the community and the rest of the energy system.

Data centers

Data centers use different ICT devices to provide the services associated with the process of storing and communicating the data behind the myriad of information services we rely upon every day. All these services are powered by electricity used by the information technology (IT) devices. On average, servers and cooling systems account for the greatest shares of electricity consumption in data centers, followed by storage drives and network devices (Shehabi et al. 2016). Given that data centers demand very large loads for the grid and are well suited for demand response programs thanks to their level of automation, there is a potential for data center participation in demand response programs (Ghatikar et al. 2012; Koronen et al. 2020).

(Smart Home) Equipment Suppliers

In-home premise equipment and appliances play an essential role in the future of demand side management. Consequently, the manufacturers of technical solutions implemented to accommodate demand response programs, also represent an important type of market actor. Some of the most interesting business opportunities for manufacturers are related to the provision of smart metering and submetering equipment, but also sensors and solutions safeguarding data and data privacy, where needed. These are required components for monitoring energy consumption at high frequency intervals to be used for information capturing and communication with, potentially multiple, services providers including retailers, aggregators, and ESCOs. Appliances for domestic, commercial, and industrial consumers increase the potential gains from their participation in demand response programs; the provision of this component (including those linked to ESCOs and/or aggregators) is also a potential vein of business for manufactures.

Distributed generation contribution to balancing services

When optimizing system operations, the need for balancing resources is strongly influenced by fluctuations in power generation and is expected to increase in future power systems due to the increased penetration of renewable power generation. Concretely, renewable energy, and especially newly installed wind power, have prompted additional demand for reserve and response operations. This demand arose predominantly from the uncertainty of day-ahead forecasts for renewable feed-ins. To appropriately respond to increased uncertainty, decentralized energy systems arise as a new player. A decentralized energy system is a relatively new approach in the power industry in most countries. Traditionally, the power industry has focused on developing large, central power stations and transmitting generation loads across long transmission and distribution lines to consumers in the region. Decentralized energy

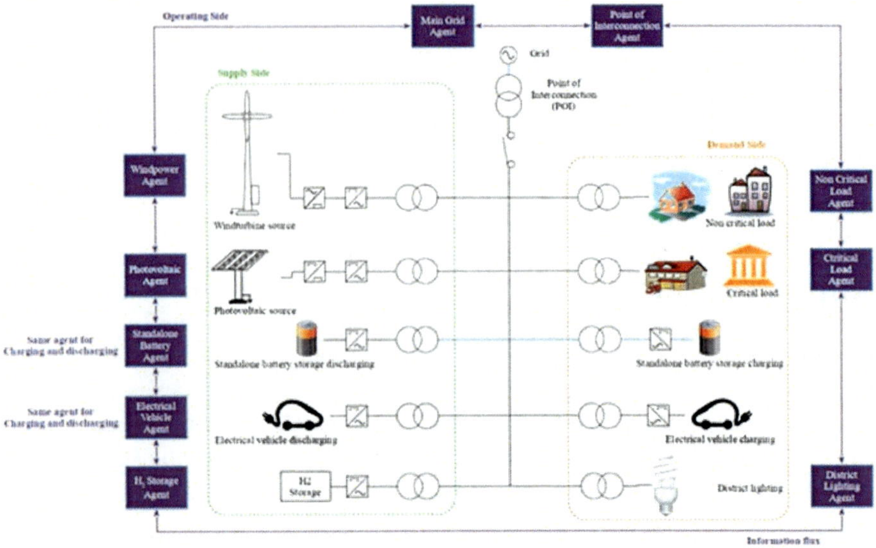

Fig. 4 Simple example (PV, wind, battery, H_2 storage, districts, lighting and POI) of interconnection and potential agents

systems seek to put power sources closer to the end user, enabling to provide power balancing services (Fig. 4).

Moving from a centralized system to a decentralized one implies that the agents work at different frequencies:

- Microseconds (for energy producers) for those who are connected to power electronics;
- Milliseconds (for energy producers) for agents managing power electronics control loops, breaker switching, voltage a frequency control, machine dynamics and transient stability, protection;
- Seconds (for energy producers), if they control batteries, super-capacitors, or mechanical system;
- Minutes (for energy producers, TSO/DSO and local system managers), for tie line regulation, global electricity load (a district for example) and hydropower plants;
- More than minutes (for energy producers, TSO/DSO, local system managers, and in general end users), for all the systems like heat storage, building inertia, heat concentrated solar power plants, and biomass-fired power plants.

Each agent has its local behavior resulting from its status, its knowledge, and its sharing of information with all the other agents of the local grid and one can also add to the agent a classical controller (model predictive control for example). Then, the agent can switch from one, the other, or both controllers depending on the following constraints of grid/market and/or characteristics given summarized below:

- Power flow control between each element of the net;
- Real-time monitoring and control;
- Effective information visualization;

- Dynamical optimization of the performance and robustness of the system;
- Quickly react to disturbances in such a way as to minimize impact;
- Effectively restore the system to a stable operating region following a disturbance;
- Fast isolation & sectionalization if a defect appears somewhere in the grid;
- Adaptive islanding;
- Anticipation of disruptive events which will lead to a blackout;
- Electricity price (spot market, etc.);
- A competitive solution compared with solutions combining variable output renewables with electrochemical storage.

4 Change in User Behavior by Smart Meter Feedback

There are two ways in which user behavior (and thus energy demand) can be affected by Smart Meter feedback, namely direct and indirect feedback (Darby 2006). *Direct feedback* can be provided via a display on the power meter itself or on a separate display. The design of the display can have a significant impact on the response—e.g., in terms of the type of information provided (qualitative or quantitative, analog or digital, etc.), the time coverage (historical, actual, expected future), the level of disaggregation, and benchmark comparisons, e.g., against previous consumption or that of some neighbors/peers (cf. Anderson and White 2009). *Indirect feedback* refers to rehashed data (e.g., consumption trends) or comparison against recommendations/target settings. Research along these lines has found that persistence of the behavioral change is higher for extended periods of information provision (greater than three months) and more detailed information (e.g., Henryson et al. 2000), and that intrinsic motivation seems more important than extrinsic one.

Demand can be influenced in various ways, e.g., by the shut-off of the energy-using device, less frequent or intensive usage, more sensible usage, or by performance (energy efficiency) improvement of the device (which often implies replacement with a more energy-efficient, state-of-the-art device) (see Table 1). Figure 5 shows different types of feedback that can be thought of by smart meter intervention (Watson et al. 2006).

5 Deep Reinforcement Learning and Building Optimization: Towards Smart Demand Response with Big Data

With the diffusion of advanced metering infrastructure, large data volumes are becoming available which can be used for the online optimization of schedules for building management systems (Mocanu et al. 2019).

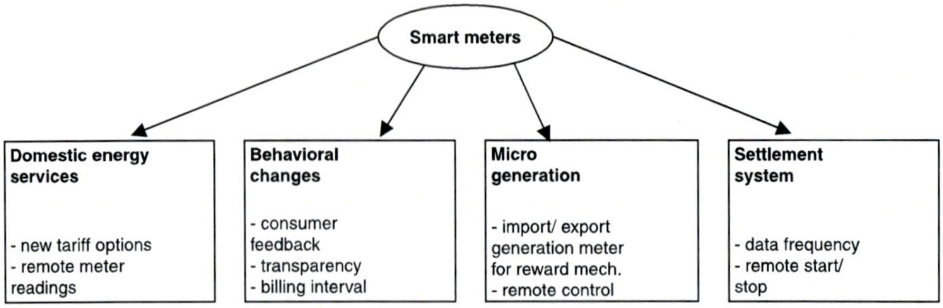

Fig. 5 Types of impact of smart metering on user behavior

Big data enables a deeper analysis and understanding of individual consumption behavior, by automatically extracting, controlling, and optimizing demand patterns, and thus bears the potential to enhance energy efficiency and optimize energy consumption in buildings (Mocanu et al. 2019).

Still, due to the curse of dimensionality, reinforcement learning algorithms/ approaches often fail in tackling large-scale problems but can be combined with others (e.g., combining deep learning with reinforcement learning). Overall, this trend has been referred to as Smart Home Energy Management Systems (in contrast to conventional HEMS).

Table 1 Five main categories of energy feedback (Darby 2006)

Type of feedback	Process	Examples
Direct	Available on demand	• Self-meter reading
		• Direct displays
		• Interactive Feedback
		• Pay-as you go meters
		• Ambient devices
		• Cost plugs on appliances
Indirect	Data processed by utility and sent to the customer	• More frequent billing
		• Comparative or historical feedback
Inadvertent	Learning by association	• Micro-generation
		• Community projects
Utility-controlled	Learning about the customers	• Smart meters
Energy audits	Learning about the energy capital of a building	• Undertaken by a surveyor engaged by the client
		• Undertaken as part of a survey
		• Carried out on an informal basis

Such hybrid machine learning (ML) approaches enable to optimize energy consumption at both the individual building and the aggregated level (with only a single agent). In combination with dynamic pricing such approaches can be used, e.g., to either pursue cost minimization or flattening the net energy demand/load profile.

6 Business Models for Smart Demand Side Management

Among the new business models for smart DSM are platform-based models involving some of the new market players described above (and in more detail in Chap. 2). Platform-based business models for DR are likely multi-lateral, i.e., involving different types of actors. A major challenge is that the notion of the customer is blurred and that there is a need to orchestrate multiple value propositions.

7 Policy and Regulatory Aspects Related to DR

Regulatory and policy aspects (e.g., tax and subsidy programs) also have an influence on the DR potentials that can be exploited.

Full exploitation of the (growing) DSM potentials due to the smart grids evolution also necessitates continuous adjustments to the legal framework and supporting policies. Smart Energy Demand Coalition (2015) has performed a review of the incentive-based DR developments in Europe. Specifically, they use the following four criteria: (1) consumer access to DR aggregators or service providers; (2) adaptation of balancing market requirements to allow for DR participation; (3) existence of standards for measurement and verification; and (4) establishment of appropriate DR remuneration schemes and penalties for non-performance, including standards for transparency and reporting (Fig. 6).

8 New Perspectives for Demand Response

Demand response becomes more valuable in light of the increasing demand for flexibility by renewable power generation. At the same time, DR, also for small consumers, becomes more accessible through ICT. Furthermore, electricity will be more and more used for heating and transportation, entailing flexibility by thermal storage units and vehicle batteries (Madlener and Ruhnau 2021).

ICT enables bi-directional communication between the grid operator/s and the end-users and thus greatly facilitates dynamic pricing, requests/offers for load flexibility, load scheduling, and availability of DER (e.g., for reducing peak demand or providing ancillary services). Smart grid enhanced DSM, on the one hand, can thus greatly help to stabilize the system, but on the other hand cause extra costs and

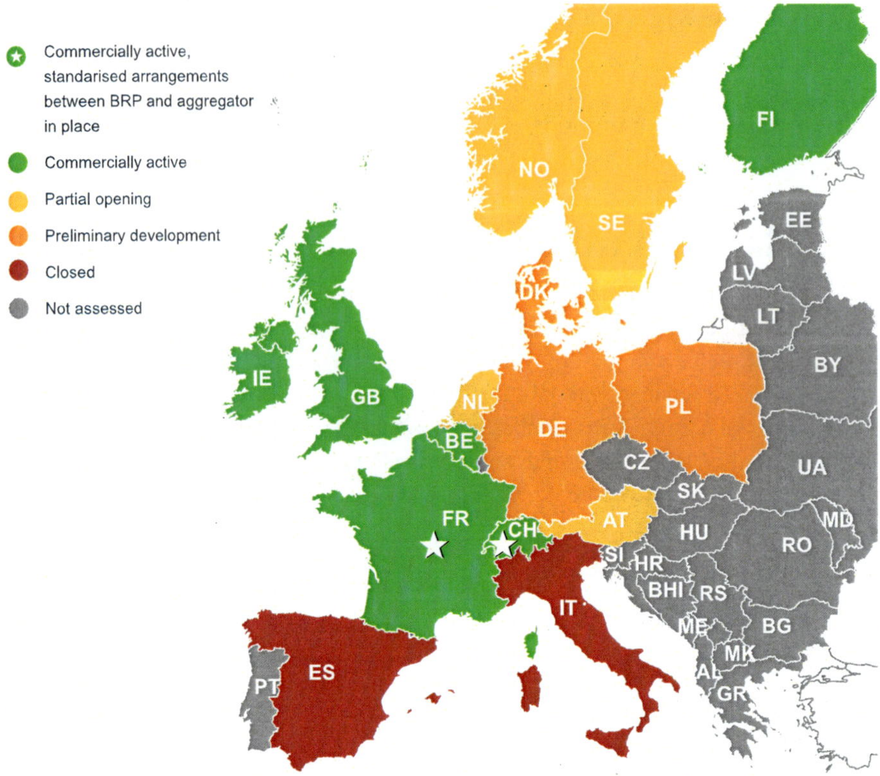

Fig. 6 Map of explicit demand response development in Europe (Smart Energy Demand Coalition 2015, p. 10)

hazards due to the impacts of bidirectional energy flows for which the system was originally not designed (Fig. 7).

In principle, two strategies can be distinguished for using DR in combination with DA and intra-day market trading: (1) activation of DA market (valued at the DA price) and (2) activation of intra-day market (valued at the intra-day price). A mix of these two strategies is also conceivable, i.e., trading DR capacity in both markets. Likewise, DR could be traded multiple times during the continuous intraday trading period (Madlener and Ruhnau 2021). DR benefits from market price volatility, which, in turn, is driven by the intermittency of renewables as well as their forecast errors.

Figure 3 shows the results from a stylized example for Germany. The assumption is that 1 kWh of electricity demand is shifted every day from the time of the day with the highest price to that of the lowest price. It is assumed that DR is available throughout but can only be activated once a day. Perfect price foresight on the energy markets is also assumed. The results show that the market value of DR is driven substantially by price volatility and that it is twice as high for quarter-hourly products than for hourly ones. With decreasing lead times it increases modestly. Note that for the case

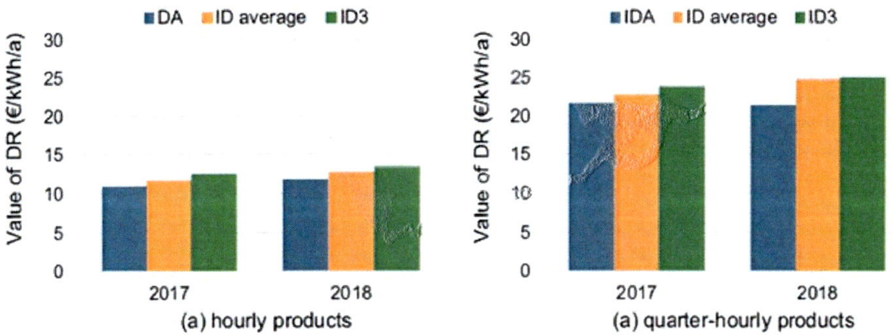

Fig. 7 The value of DR in the German electricity spot market 2017–2018. DR activation strategies are compared for hourly **a** and quarter-hourly **b** products in the DA and intra-day auction as well as in continuous intra-day trading, using the volume-weighted average intra-day price over the entire trading period (intra-day average) and over the last 3 h of trading (Madlener and Ruhnau 2021)

of capturing the (higher) price volatility of quarter-hour products, it is necessary that the DR application must be able to adjust demand by 4 kW, rather than by 1 kW only for the hourly products (Madlener and Ruhnau 2021).

The role of DR being integrated into a portfolio of renewable power generation assets, jointly optimized by trading in the intraday and day-ahead markets, has been studied by (Garnier and Madlener 2016). They show that the day-ahead activation of DR can increase the sales revenues of renewables, and thus the day-ahead market value, whereas the intra-day activation of DR can help to reduce the balancing costs of renewables.

9 Conclusions and Final Remarks

In this section, we have first provided a short overview of the various flexibility options to accommodate growing shares of variable renewables, including DR, followed by a discussion of the major forms of DR programs and the various challenges involved. The issues brought up are generally relevant for residential, commercial, and industrial end-users alike, although the practical specifics might be quite different from each other.

Second, the role of dynamic pricing and tariffs is discussed as an important lever for activating and enhancing consumer participation in DSM, alongside the various DR schemes available. The special role of ESCOs in this context is emphasized (here primarily in a conventional energy system setting).

In the third part, the different new market actors and emerging business opportunities are covered, especially ESCOs (now in the context of the evolution of smart grids). In particular, the role of aggregators, ESCOs, smart appliance equipment providers, and "Energy Suppliers 2.0" were addressed, as well as the importance of (reducing) transaction costs.

Finally, we addressed the need for policy action and regulatory change to support cost-effective demand response.

Overall, we pointed out that while DSM/DR has been used for decades, the evolution of the smart grid and the emergence of new actors bear an enormous potential for better exploiting the DSM potentials for balancing energy supply from increasing shares of variable energy resources (VER) with the rapidly changing needs (and thus load profiles) of both consumer and prosumer households. DSM is one of several options for flexibility. It needs to be seen how the "merit order of flexibility options" evolves over time.

10 Case Study 1: Demand Side Management Experiences

The ongoing power sector transformation is being accelerated by the combination of electrification, decentralization, and digitalization. The application of digital monitoring and control technologies in the power generation and transmission network has been an important trend for several decades, and has recently started penetrating deeper into power systems. Wider usage of smart meters and sensors, the application of the Internet of things (IoT) and the use of large amounts of data with artifical intelligence (AI) have created opportunities to provide new services to the system and the emergence of new business models.

There are multiple examples of new business models associated with this transformation process. More specifically, in this case study, some illustrative examples of system flexibility provision, will be discussed. Through ICT and digital technologies, all industrial and residential consumers can provide flexibility services.

This trend, driven by the increasing deployment of smart grids, smart meters and other software platforms that capture data on energy use, is expected to rapidly gain ground in the coming years and become a more mainstream way for businesses to reduce energy consumption and costs.

According to research on the future of the DSM market published by Navigant Research,[1] emerging DSM technologies are starting to offer energy users better ways to reduce energy costs by increasing their focus on behavioral measures and data analytics. This is a promising future, where according to this report, DSM and new business models are expected to lead to an estimated US$63.6 billion in worldwide spending in 2028.

In order to benefit from this increasing market, a range of solutions providing the new energy services are being developed based on automation and market signals. In the case of the domestic and small and medium enterprises segment, these solutions can be grouped in the following fields:

Heating, Ventilation and Air-Conditioning

[1] https://guidehouseinsights.com/reports/demand-side-management-overview.

Since HVAC systems represent a significant portion of domestic energy consumption, its potential participation in DSR mechanisms has gained attention in recent years. This highlights the importance of developing building energy control and operation strategies, which would help to increase the energy efficiency and indoor comfort in the buildings. For this purpose, energy-modeling tools have been developed, with the aim of minimizing the energy cost while respecting the customer comfort.

When providing these DSR solutions, independent aggregators have been active in offering technology to residential consumers in order to connect appliances so as to aggregate their flexibilities. For instance, one European aggregator uses a wireless transmitter and electricity modulator provided to consumers, installed and operated without charges, to connect appliances such as electrical heaters, air conditioners, heat pumps and water boilers in homes, commercial buildings and offices. The aggregated flexibilities are sold to the grid operator for balancing and safety purposes, and on a daily basis to all players through wholesale markets. The cooperation with manufacturers of heating devices to directly include control technology into the devices, enables consumers to participate in DR offers without further investments costs for the consumer or the aggregator. This enables the current heating system to operate efficiently and helps save up to 15% of heating expenses (Eurelectric 2017).

Home Battery Automation

Increasing adoption of EVs fosters the development of new business models associated with home battery automation. These Energy as a Sevice (EaaS) models could provide storage systems for customers to store energy during periods of low demand and to draw from that stored energy during periods of peak demand.

In some European countries, home battery automation has been advanced and commercially developed by electricity suppliers or energy services companies. In exchange for a discounted home-battery, the consumer agrees to let them use a percentage of the battery capacity. The aggregated capacity is then used to spread over its consumer portfolio to create a virtual power plant providing balancing services to the grid.

Transactive energy management solutions

The market for energy management solutions has also been developing on pace. Under this business model, an aggregator accumulates transactive loads such as electric thermal storage heaters, hot water heaters, ice-based air conditioning, compressed air energy storage, and electric vehicles into a grid asset (Virtual Power Plant), to deliver ancillary services to grid operators. They also monitor real time price variation and purchase electricity when it is cheapest. Through the combination of savings from arbitrage with profits from ancillary services, they provide low-cost ancillary services to grid operators and reduce the cost of heating for consumers.

The provision of this demand side flexibility, defined as the ability of a customer to deviate from the normal electricity consumption in response to economic signals, could be monetized through the active participation of the customer in the power system operation and its associated markets (Table 2) providing services to the TSO, the DSO or the actors participating in these markets and with the BRP.

Table 2 Different roles of demand side flexibility

	Service provided to the power system	Potential client		
		DSO	TSO	BRP
Constraint management	Voltage control	X	X	
	Grid capacity management	X	X	
	Congestion management		X	
Adequacy	Capacity payment		X	X
	Strategic reserve		X	X
	Hedging		X	X
Wholesale market	Day ahead optimization		X	X
	Intraday markets		X	X
	Generation optimization	X	X	
Balancing markets	Frequency containment	X	X	
	Automatic and manual frequency restoration	X	X	
	Replacement reserve		X	X

11 Case Study 2: Market Design for Greater System Flexibility

The increasing share of RES deployment required to reach the sustainability targets stated by the European Union will have increasing impacts on the functioning of electricity markets. When replacing conventional power plants with renewables such as wind turbines and solar PV, the ability to provide power balancing services in the classical sense disappears. As the intermittent electricity generation from fluctuating renewables increases, the need for balancing services will consequently also increase. Furthermore, conventional fossil fuel power plant generators are synchronous with the grid, and therefore, provide rotating inertia that supports the system frequency against changes.

It is, therefore, evident that the transition towards a decarbonized power system will lead to challenges of balancing the electricity supply and demand. Alternative sources of balancing services must therefore be established as the conventional power plants are pushed out. One of the approaches to obtaining alternative balancing services is the smart grid concept, where flexible consumption takes part in the balancing effort. Demand side response could be the main contributor to more effective markets and to system security with a high penetration of fluctuating generation.

Historically, generators, such as gas-fired power plants and pumped hydropower plants, have mainly supplied flexibility. Nevertheless, in an increasing renewable generation scenario, this could not be enough. Demand-side resources can provide

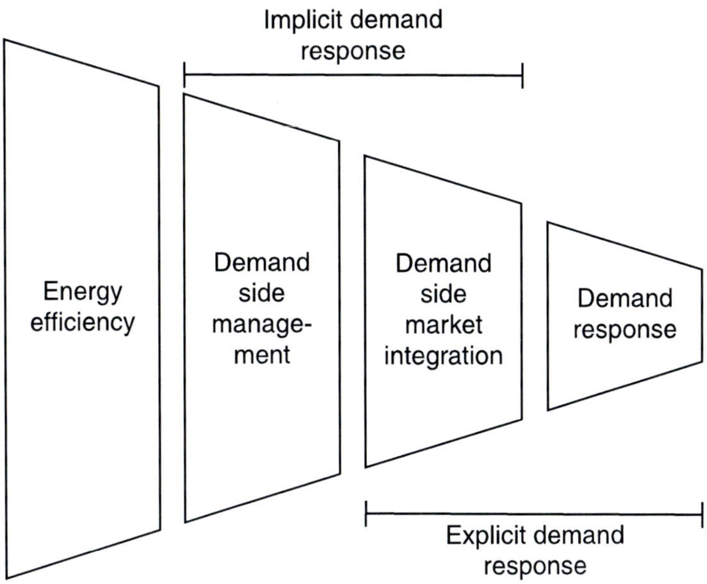

Fig. 8 Demand side engagement (ACER 2017)

flexibility through demand response. Industrial consumers are already offering flexibility this way to the system operator, by agreeing to reduce or increase consumption as needed, in return for compensation. Residential or small commercial consumers can also provide demand response, but the relatively small scale of the flexible loads of these consumers requires new approaches to provide these flexibility requirements through new energy actors such as aggregators.

Different kinds of demand side activities have been considered worldwide for demand side participation. Focusing on explicit (industrial and residential consumers adapting their consumption patterns to price without explicitly participation in the market) and implicit demand response services (demand response is explicitly sold by consumers to the market or to grid operators), the market design has to be adapted accordingly for enabling both types of DR (Fig. 8).

The envisaged transformation is likely to require numerous regulatory adjustments and the implementation of new market design elements to efficiently respond to the expected changes coming both from the supply and the demand side. That requires common rules and standards for the operation of power grids with different approaches.

European Experience

The recently approved Clean Energy Package includes a set of measures to push forward the energy transition with the consumers at the center of this process. Integration of demand side flexibility has been identified as an important component of the European Union's transition towards a low carbon economy. This case study will provide some insights on the most relevant aspects that provide new business opportunities.

The development of such rules has been initiated by the Agency for the Cooperation of Energy Regulators (ACER), establishing guidelines to which these common rules and standards need to comply with. Based on these guidelines, the European Network of Transmission System Operators for Electricity (ENTSO-E) drafted eight network codes with the aim to enable the implementation of the EU internal electricity market. These network codes have the legal status of directives. The network codes on Capacity Allocation and Congestion Management (CACM) and Electricity Balancing (EB) concern the design and operation of short-term electricity markets across EU Member States and thus provide a framework under which potential developments of these markets can be materialized. Demand side response is considered a key component of future power system operation where residential consumers offer potential for load balancing as well as economic savings.

US experience

In the United States, according to North American Electric Reliability Corporation (NERC), higher penetration of variable generation resources has created a growing need for flexible resources, such as demand response, in order to balance electricity supply and demand, ensure resource adequacy, and meet ramping needs. Planners and operators could face challenges when integrating variable generation resources and other emerging technologies as inputs, potentially requiring revisions to operational practices, enhancement of reliability standards, and changes in market designs.

The rise of renewables is expected to continue, at least for the next few years. In this regard, some electricity markets, such as the California Independent System Operator (CAISO) has started to recognize, to varying degrees, flexible and resilient electric resources. Likewise, policy makers at the Federal Energy Regulatory Commission (FERC) and the PJM Interconnection are shifting focus, in the United States at least, to the role that battery energy storage and flexible resources like distributed resource aggregators (DRA) could play as electricity markets evolve.

In the field of DER, they have made their contribution; (smart grid supported) DER is an emerging resource category too often overlooked, with aggregation pilots focused on monitoring and offering services to the grid.

Asia-Pacific experience

Asia-Pacific's liberalized electricity sector is attracting active participation from private companies and customers in the DR market. Countries such as Japan and Singapore have already developed DR regulations and are in the process of testing DR to assess its viability and arrive at the right mix of incentives to encourage private participation.

As a large industrial nation with a forecasted peak load of more than one Terawatt by 2025, China is a potentially large market for demand response management systems. The country has experienced shortages in power availability because of rapid economic growth, a situation that has been solved in recent years. Nevertheless, to manage this mismatch between electricity demand and supply, large industrial customers are one of the most relevant players in these DR programs focused on reducing peak demand.

Overall, demand side response is considered a key component of future power system operation where residential consumers offer potential for load balancing as well as economic savings.

Review Questions

- Can you identify and describe the five main flexibility options for accommodating increasing/high shares of renewables?
- Are you able to list five main barriers for the wider uptake of demand response options?
- Can you name and explain three different types of DR programs, and the most important differences between them?
- Are you able to explain the role of aggregators in managing distributed energy resources?
- Can you describe the main criteria for (demand-side) flexibility market design?
- Why are energy monitoring platforms attractive for ESCOs?
- Sketch the demand side engagement funnel. Can you explain its main elements?

References

ACER (2017) Encouraging flexibility in electricity market design

Aghaei J, Alizadeh M (2013) Demand response in smart electricity grids equipped with renewable energy sources: a review. Renew Sustain Energy Rev 18:64–72

Albadi M, El-Saadany E (2008) A summary of demand response in electricity markets. Elect Power Syst Res 78(11):1989–1996

Albadi MH, El-Saadany EF (2007) Demand response in electricity markets: an overview. In: 2007 IEEE power engineering society general meeting. IEEE, pp 1–5

Alizadeh M, Li X, Wang Z, Scaglione A, Melton R (2012) Demand side management in the smart grid: information processing for the power switch. IEEE Signal Process Mag 29(5):55–67

Anderson W, White V (2009) Exploring consumer preferences for home energy display functionality. Rep Energy Saving Trust 123:1–49

Batalla-Bejerano J, Trujillo-Baute E, Villa-Arrieta M (2020) Smart meters and consumer behaviour: insights from empirical literature. Energy Policy 144:111610

Borenstein S, Jaske M, Rosenfeld A (2002) Dynamic pricing, advanced metering, and demand response in electricity markets. Energy Series, The Hewlett Foundation

Castillo-Cagigal M, Caamaño-Martín E, Gutiérrez A, Bote DM, Monasterio-Huelin F, Porro J, Matallanas E, Jimenez-Leube FJ (2010) Self-consumption of pv electricity with active demand side management: the gedelos-pv system. In: Proceedings of the fifth world conferences on photovoltaic. Energy Conversion, pp 4866–4870

Darby S (2006) The effectiveness of feedback on energy consumption. Technical report, Oxford, UK

Darby S (2010) Smart metering: what potential for householder engagement? Build Res Inf 38:442–457

Denholm P et al (2010) The role of energy storage with renewable electricity generation. US National Renewable Energy Laboratory
European Commission (2016) Evaluation of the eu framework for metering and billing of energy consumption. EC, Brussels
Faria P, Vale Z (2011) Demand response in electrical energy supply: an optimal real time pricing approach. Energy 36(8):5374–5384
Garnier E, Madlener R (2016) The value of ICT platform investments within distributed energy systems. Energy J 37(SI2):145–160
Ghatikar G, Ganti V, Matson N, P M (2012) Demand response opportunities and enabling technologies for data centers: findings from field studies. Technical report
Hargreaves T, Nye M, Burgess J (2010) Making energy visible: a qualitative field study of how householders interact with feedback from smart energy monitors. Energy Policy 38:6111–6119
Henryson J, Håkansson T, Pyrko J (2000) Energy efficiency in buildings through information-Swedish perspective. Energy Policy 28(3):169–180
Hinnells M (2008) Technologies to achieve demand reduction and microgeneration in buildings. Energy Policy 36:4427
IEA (2019) World Energy Outlook 2019. International Energy Agency
IRENA (2018) Power system flexibility for the energy transition. International Renewable Energy Agency
Koronen C, Ahman M, N L J (2020) Data centres in future Europeean energy systems–energy efficiency, integration and policy. Energy Effic 13:129–144
Madlener R, Ruhnau O (2021) Chapter 13—variable renewables and demand flexibility: Day-ahead versus intraday valuation. In: Sioshansi F (ed) Variable Generation, Flexible Demand. Academic Press, pp 309–327
Mocanu E, Mocanu DC, Nguyen PH, Liotta A, Webber ME, Gibescu M, Slootweg JG (2019) On-line building energy optimization using deep reinforcement learning. IEEE Trans Smart Grid 10(4):3698–3708
Mohagheghi S, Stoupis J, Wang Z, Li Z, Kazemzadeh H (2020) Demand response architecture: Integration into the distribution management system. In: Proceedings of the IEEE smart grid communications. Gaithersurg, MA, USA, pp 501–506
Morstyn T, Farrell N, Darby SJ, McCulloch MD (2018) Using peer-to-peer energy-trading platforms to incentivize prosumers to form federated power plants. Nat Energy 3(2):94–101
Osterwalder A, Pigneur Y (2010) Business model generation. A handbook for visionaries, game changers, and challengers. Wiley
Salah F, Flath CM, Schuller A, Will C, Weinhardt C (2017) Morphological analysis of energy services: paving the way to quality differentiation in the power sector. Energy Policy 106:614–624
Shehabi A, Smith S, Horner N, Azevedo I, Brown R, Koomey J, Masanet E, Sartor D, Herrlin M, Lintner W (2016) United states data center energy usage report. Lawrence Berkeley National Laboratory, Berkeley, California. LBNL-1005775 Page, vol 4
Smart Energy Demand Coalition (2015) Mapping demand response in Europe today. Technical report
Specht JM, Madlener R (2019) Energy supplier 2.0: a conceptual business model for energy suppliers aggregating flexible distributed assets and policy issues raised. Energy Policy 135:110911
Vine D, Buys L, Morris P (2013) The effectiveness of energy feedback for conservation and peak demand: a literature review. Open J Energy Effic 2 (2013)
Watson J, Sauter R, Bahaj B et al (2006) Technical report. SPRU, Brighton

Market Engineering for the Smart Grid

Philipp Staudt and Christof Weinhardt

> **Learning Objectives**
>
> - What is Market Engineering?
> - How do Smart Girds impact the energy market?
> - How can Market Engineering and the Smart Grid improve energy markets?

1 Energy Market Design in a Changing Environment

In January 2001, California experienced rolling blackouts in their energy systems and an average price of electricity of $250 per MWh. This price was nearly ten times the average price of the previous January in the year 2000 (Woo et al. 2003). What had happened? California had liberalized its electricity market with a zonal setup and with it, had introduced various new market components that were intended to reduce grid congestion, market power and risks for consumers, and at the same time, drive down wholesale electricity prices. There was little experience with designing electricity markets at that time and some market design choices created strategic incentives for individuals to optimize themselves against the market (see Alaywan et al. (2004) for more details on California's design flaws). California was not the last trial and error of power market design. There are constant reports on small power market failures all over the world. This is not surprising as the market design for

P. Staudt · C. Weinhardt (✉)
Institute of Information Systems and Marketing, Karlsruhe Institute of Technology (KIT), Kaiserstraße 89-93, 76133 Karlsruhe, Germany
e-mail: christof.weinhardt@kit.edu

P. Staudt
e-mail: philipp.staudt@kit.edu

C. Dinther et al. (eds.), *Smart Grid Economics and Management*,
Lecture Notes in Energy 51, https://doi.org/10.1007/978-3-030-84286-4_4

power markets is very complex and it needs to integrate various interests. The peculiarities of the power market have been introduced to you in chapter "Smart Grid Economics". The worldwide transition to more intermittent renewable generation challenges traditional market designs. It leads to generation spikes that can cause congestion (Staudt 2019), reversed power flows on the distribution level (Walling et al. 2008) or even negative electricity spot prices (Kyritsis et al. 2017). The market design needs to be adapted to these changes. At the same time, the Smart Grid introduces new possibilities to measure and control actions at the low voltage level that can help to balance the local infeed of renewable generation. This allows new actors to enter the market, creates active and price-sensitive consumers and provides more detailed market information. It is important to note that when we talk about the power market (in this Chapter we will use power, electricity or energy market synonymously), we are really meaning a multitude of market stages (i.e., sub-markets) such as the wholesale spot and intraday markets or the market for balancing power, for example. Others might be added in the course of the energy transition, such as redispatch markets (Hirth and Glismann 2018). The Smart Grid potentially impacts all of these sub-markets and it might enable the creation of new sub-markets, for example, in the form of peer-to-peer trading in the distribution grid (Mengelkamp et al. 2018). Such markets as well as changes in the existing markets need to be carefully engineered to avoid market failures as in California and to achieve the intended objectives. In this chapter, we therefore, introduce the market engineering framework as a way to systematically describe and engineer existing and new markets. We then describe the impact of the Smart Grid on energy markets and discuss how these changes can be classified.

2 Concept of Market Engineering

As shown by the example of the failed Californian electricity market reform, a market engineer needs to consider a variety of aspects when designing a market that also needs to be continually re-evaluated. A formalization of the components of market engineering in the form of a market engineering framework is provided by Weinhardt et al. (2003a). The designed framework is depicted below in Fig. 1. The framework is originally designed to describe virtual marketplaces but it can be used to formalize all kinds of markets. It is a very powerful tool that forces the engineer to consider different aspects of a market environment when designing the market. In the following, we describe all components of the framework. Note, that the market engineer can only influence the market structure that is comprised of the microstructure, the (IT)infrastructure and the business structure. However, if the market engineer is the regulator or the designer of the traded product, she might also be able to influence the transaction object that is traded on the market. After the explanation of the framework, we provide an illustrative example of the secondary control power market in Germany. Additionally, we describe some desired properties of markets that should be taken under consideration when the market is designed.

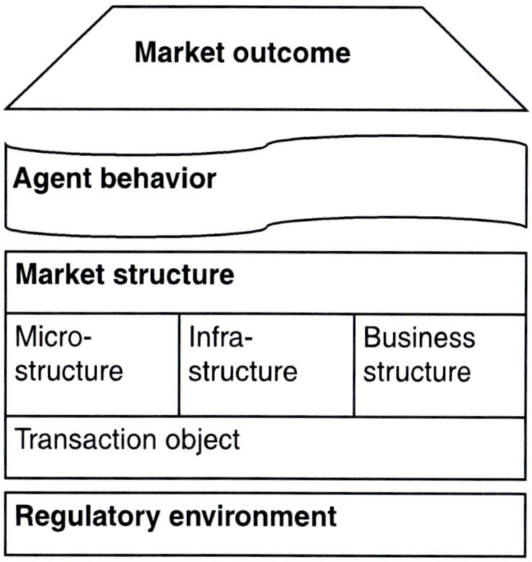

Fig. 1 Market Engineering Framework (Weinhardt et al. 2003b)

2.1 *The Market Engineering Framework*

Socio-economic and legal environment
This is the foundation for the market engineer. The rules of the market, the traded products, and the participating agents need to comply with laws and regulations. This might include tax law, competition law and so on and so forth. In the electricity market context, there might be specific regulations on the availability of resources or specific limits for the allowed emissions of different substances, for instance.

Transaction object
The transaction object is the product that is to be traded on the designed market. It might be abstract such as any legally tradable physical good as would be the case for eBay, for instance. In the electricity context, it would often be either power or electricity. However, with the Smart Grid, new demand side flexibility resources might become tradeable. A possible product formulation is introduced by Dauer (2016).

Microstructure
The microstructure determines the rules of the market. This can be broadly summarized as the auction design on the designed market. It defines how sellers and buyers submit bids and how a successful trade is found and at what price it is executed. Even a farmers market has a microstructure. Customers that arrive at the market before the produce is completely sold, can perform a successful transaction if they are willing to pay the displayed price.

IT-infrastructure
Today, most markets are virtual. Market places can be accessed on web platforms and bids can be submitted online. This means that markets need an IT infrastructure that determines the access for different participants, the time resolution and needs a

user- friendly design. eBay, for instance, allows access to all registered users within the timeframe of the auction to access and to enter their reservation price using a number field.

Business structure

The business structure is the business model of the market operator. It determines her revenue model. For example, participants might pay an access fee and a fraction of each transaction to the market operator. If the market operator is not a private company but the regulator, the business structure relates not to revenue but some other metric that is to be achieved. One example is the resiliency of the electrical grid that is strengthened through the procurement of reserve power.

Agent behaviour

The microstructure, the IT infrastructure and the business structure form the overall market structure. It can be influenced by the market engineer to achieve eventual goals of the market, be it efficiency, reliability or profit maximization, among many others. However, the structure does not directly translate into a market outcome. It defines a set of rules and regulations that influence agents on the market. The market result then depends on the behaviour of agents that act within this set of rules. For instance, a very complicated framework might discourage participation and therefore, the ultimate outcome cannot be achieved. The market structure of eBay, for instance, encouraged agents to submit their bids last second in order to impede other people from being able to outbid them. This even led to according software artefacts that would allow people to follow this strategy more conveniently.

Market outcome

The market outcome results from the agent behaviour. Usually, the market engineer has a set of objectives for the market outcome as previously described. These objectives can then be evaluated against the actual outcome. For instance, the German government has introduced a market mechanism for the assignment of feed-in tariffs to large renewable generation units. They are assigned through an auction that is performed multiple times per year. The objective is to reach the expansion goals for renewable generation while using the cheapest technology at the most suitable locations in terms of total generation. In the year 2019, almost no new wind projects were proposed, which is supposedly caused by increasing opposition within the population and other regulatory issues. As this is not caused by the market design, which is suitable, the government is currently evaluating a change to the legal environment. This is possible as the government has a further reach than normal market engineers. The performance of a market is difficult to measure because it depends on the desired properties of the market. Therefore, we introduce possible performance critieria of a market in Sect. 2.2.

Case study: Market design of the German secondary reserve market

To illustrate, how the market engineering framework can be used to describe and design markets, we provide the example of the secondary control energy market in Germany. This market is operated by the TSOs in Germany, who need to procure reserve power in order to balance the grid in case of unforeseen events or forecasting errors and it serves well as an example because its setup is rather complex. The secondary reserve is activated after the primary reserve power. The costs for the power provision are covered through grid tariffs by all consumers. The actual cost of activating the secondary reserve is distributed among the market participants that caused the disruption of the balanced grid operation. The socio-economic and legal environment for reserve power is constituted by the law on grid access (in German *Stromnetzzugangsverordnung (StromNZW)*). There are of course many other laws to consider, for example, the law on energy markets (*Energiewirtschaftsgesetz (EnWG)*). However, it often suffices to describe only the most important regulation. The transaction object on the secondary reserve power markets is multi-dimensional. It is composed of a quantity, a power and an energy price, it can be positive or negative (meaning that power can be added to or taken from the system), and it is differentiated by the time of day in four-hour blocks. The quantity is the offered amount of power that can be activated to balance supply and demand in the energy system. The power price dimension provides the payment that is required to be paid to contract that offer. It is paid regardless of an activation to provide balancing power. The energy price, on the other hand, is only paid if the participant is actually required to provide energy. Positive and negative secondary reserve power is procured separately. Furthermore, the products are divided into four-hour blocks in which they have to be provided. Therefore, the transaction object has 5 dimensions. The microstructure has multiple stages: in the first stage, the TSOs acquire reserve capacity and in the second stage they procure reserve energy. The capacity stage serves as a form of insurance, ensuring that enough reserve energy is available. Reserve energy can be thought of as a transaction object with a capacity price bid of zero. On the reserve capacity market, the bids are accepted in increasing order until the required capacity is reached. Then, in the reserve energy market, the participants that have been accepted at the reserve capacity stage have to participate, but other market participants are also allowed to submit bids. The required reserve energy is then also acquired in increasing order. The TSOs use a web interface as IT infrastructure. The business structure for the TSOs is to generate a competitive environment and to procure the secondary reserve at a minimal price. They, therefore, do not have a revenue objective but they do have a financial objective. In the past, the agent behaviour was often strongly influenced by the market structure, because the bids for reserve power and energy were submitted simultaneously and accepted only based on the capacity price. Therefore, there was no more competition for the energy price. This lead to situations where suppliers bid very low power prices but the energy prices skyrocketed, which then occasionally lead to very high reserve energy prices. Consequently, this mechanism was redesigned. This is a very good example of how market engineers react to the market outcome by trying to change the agent behaviour through a redesign of the market structure. As the market design changed only very recently, it remains to be seen whether it has an effect on the market outcome.

2.2 Market Properties

There is a variety of market properties and quality measures that can be considered when markets are designed. In the previous example, the objective of the market operators is to achieve minimal costs because they are both, the market operators and the consumers on the market. Often, a market engineer needs to consider the interests of both, supply and demand on a market, which results in properties that take all stakeholders into account. These properties are sometimes conflicting, sometimes they are complementary. The following concepts are extracted from Wurman (1999, pp. 15–19) as it provides a neat overview but many concepts are originally explained by Mas-Colell et al. (1995).

- **Efficiency**: There are different definitions of efficiency. One could, for example, argue that an efficient solution is such that it maximizes the utility of all agents. However, a more common definition of efficiency is Pareto efficiency. A market leads to a Pareto efficient solution if in that solution, no agent can improve her utility without reducing the utility of another agent.
- **Equilibrium**: There is a variety of different possible equilibria. However, a market engineer might wish to build a market such that some form of equilibrium state can be achieved. One common definition is that of a Nash equilibrium in which no player wishes to deviate given the rational strategy of the opponents. A dominant strategy in this sense would be a strategy that is optimal for a player regardless of the actions of her opponents.
- **Stability**: Formally, this concept means that agents will not choose a strategy that is not allowed by the designed mechanism that would increase the overall utility of the agents. A solution is, therefore, stable because no agent has any incentive to work around the mechanism to improve her position. Therefore, any stable solution is also Pareto efficient but not vice versa.
- **Individual Rationality**: A mechanism is individually rational if it is stable for each individual agent. That means that every agent would enter the mechanism and not try to work around it. In essence, it means that an agent cannot be worse off by participating in the mechanism.
- **Convergence**: This implies that if an equilibrium exists, it will eventually be reached by the mechanism.
- **Incentive Compatibility**: A mechanism is incentive compatible if it is the dominant strategy for each agent to reveal her true type. This is understood the easiest when considering an auction. Every agent has a valuation for a given good. In an incentive compatible mechanism, each agent has the incentive to report the true valuation to the auction operator. Every other strategy potentially leads to a worse payout.
- **Privacy preservation**: This simply means that the mechanism should not allow agents to learn relevant private information about other agents that could be used in future executions of the mechanism.

- **Computational Costs**: The necessary communication, the number of necessary messages and the computational complexity of associated algorithms of the mechanism should be kept to a necessary minimum.

There might be other desirable market properties that are not included here. For instance, in some markets, perceived fairness might be an important factor. However, not all of these properties do necessarily have to be addressed. However, if a mechanism has no equilibrium, for example, there needs to be some form of decision rule that determines market results. Ultimately, designing markets is an engineering task that should be pragmatically executed to achieve the desired outcome.

3 The Smart Grid and Energy Markets

The Smart Grid as such does not change the market. In essence, it is merely a layer of communication infrastructure that creates the opportunity to communicate signals in real-time or near real-time between market participants. Additionally, analyzing this data can help in making investment decisions in the system that increase its efficiency (see also chapter "Smart Grid Analytics"). Through these signals, the Smart Grid can indirectly influence demand and supply curves by allowing new actors to participate in the energy market and by allowing consumers to react to market signals. Figure 2 shows a reduced causal model for an efficient market outcome. The numbers 1 to 3 indicate the components that can be influenced through the installation of a Smart Grid. In essence, the market outcome is influenced by supply and demand curves which are efficiently integrated through a market design. Supply curves are influenced by the amount of competition in a market. More competition forces participants to bid their actual marginal costs of generation (Bompard et al. 2007). The demand curve is particularly influenced by the demand elasticity. One peculiarity of the electricity market is that demand is often very inelastic. This can lead to inefficiencies due to price caps, too few options of the demand side to participate in the market or regulation on market power abuse (Cramton and Ockenfels 2016). Therefore, increasing demand elasticity can increase the efficiency of electricity markets (Bompard et al. 2007). The Smart Grid can impact the supply side by allowing new actors to participate in the market. Small generators, storage units or electric vehicles, among others, can be aggregated and controlled through signals communicated through the Smart Grid. This increases competition, and therefore, leads to more competitive behaviour causing a more efficient market outcome. At the same time, the demand side is enabled to react to price signals through automated energy agents that operate storage capacity, heat pumps or electric vehicles. This increases price elasticity through changed time preferences in consumption. Furthermore, market information can lead to changes in the long run. Information on price development and individual consumption can cause consumers to switch their supplier contracts to time-variable tariffs and consumers might invest in storage or PV panels because they are able to estimate the effects of preferential self-consumption more easily.

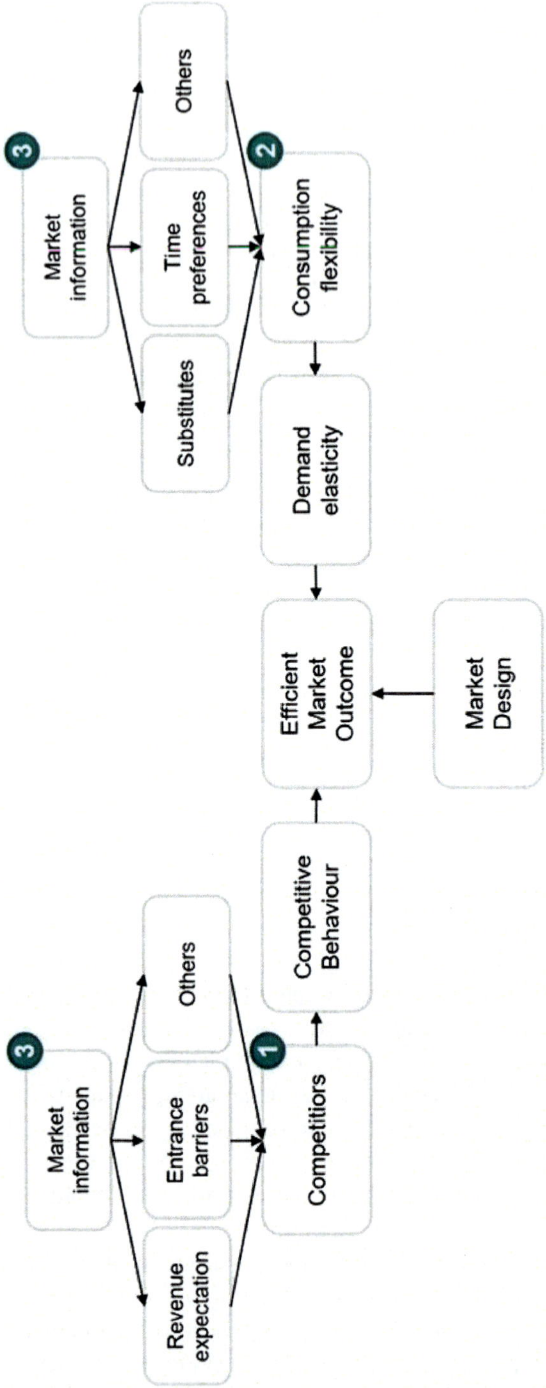

Fig. 2 Causal model of market outcome

The purpose of markets always is the coordination of supply and demand through signals. In the long run, these markets are intended to incentivize the investment in resources that can increase the efficiency of the market. For example, in the housing market, an increase in demand leads to increasing rents. This signals scarcity and might incentivize people to rent smaller apartments or to share apartments but might also cause people to sublet parts of their house as the high rents make this more attractive. This might lead to coordination mechanisms such as websites that connect people who are looking for shared apartments. The high rents make it more attractive to build and sublet apartments. This extends the offer and increases the amount of rented apartments. Information is an important factor in this process. If the level of apartment rents is not public knowledge, market participants might not be able to interpret them correctly. Similarly, in the electricity market, an increasing demand might cause increasing prices. This might lead to energy efficiency measures causing some people to be able to share their PV generation with their neighbours. Online platforms can help in finding tariffs that reward off-peak electricity use or such platform can connect neighbourhoods into microgrids that share their generation. In the medium term, this price signal would cause private households to install more PV capacity on residential roofs or to invest in storage capacity to maximize preferential self-consumption. Similarly, as for the housing market, the market signals and information need to be communicated to consumers. The Smart Grid is the infrastructure that allows all of these market signals to be efficiently communicated and similar solutions as on the housing market to be implemented for the electricity market in the distribution grid.

The intention of this chapter is to introduce changes to the market caused by the implementation of the Smart Grid. The areas that benefit from the Smart Grid can broadly be divided into the three areas highlighted in Fig. 2. These three areas can then be differentiated into market operation and market evolution where market operation describes the short-term opportunities created by the Smart Grid that affect supply and demand in operation, while market evolution encapsulates the mid- to long-term efficiency increases in the market. Therefore, in the following sections, we introduce examples of market improvements potentially caused by the Smart Grid along the value chain of generation, distribution and consumption. The examples are depicted in Table 1. We discuss the impact of the Smart Grid in particular examples and what part of the Market Engineering framework is impacted.

Table 1 Market developments triggered by smart grid developments

	Market operation		Market evolution
	Competitors	Consumption flexibility	Market information
Generation	Virtual power plants	Net metering	Hardware investment
Distribution	Congestion markets	Peak pricing	Community storage
Consumption	Smart energy communities	Real-time pricing	Tariff recommendation platforms

4 Market Operation

Market operation naturally refers to anything that can be adapted quickly and without long lead times. Such quick wins of the Smart Grid can be the activation of market participants through communication, control and the transmission of market signals. The Smart Grid is in essence the infrastructure to quickly transmit information and instructions regarding the electricity system. This information can be a price that is a signal to a generation or flexibility asset to be activated or it can simply be a price signal that causes a response of consumers. In the following sections, we first introduce examples of how the Smart Grid enables new resources to enter the market and how it enables new markets to form. This is followed by new forms of tariff designs that will increase the temporal flexibility of the demand side (see also chapter "Demand Side Management").

4.1 Competition

The increasing availability of real time data through the Smart Grid enables operators to coordinate a number of resources for different purposes. Such purposes might be frequency stability, congestion relief or the exploitation of temporal price differences. Coordinating these resources and allowing them to enter the market increases competition. While the market design might not change, the agent behaviour might change because of the increased competition. In the following, three new market developments are described which are enabled by the Smart Grid and relate to an increase in competition.

Virtual power plants. Virtual power plants are one of the major concepts enabled by the Smart Grid. The term describes a connection of different distributed energy resources (DER) that act together imitating a conventional power plant. The combination of these resources then jointly forms a virtual power plant. The resources can be anything from renewable capacity such as PV panels or wind turbines as well as active demand response, electrical storage or electric vehicles. A very good description of this concept and its characteristics is provided by Pudjianto et al. (2007). The authors distinguish between the capacity and the energy effects of DER. They argue that energy generation from DER can replace generation from conventional power plants. However, their capacity does not replace conventional capacity, thus leading to over-provision of capacity. Through combinations into virtual power plants, DER become visible from a capacity perspective and can be actively relied upon by system operators. If you think back to the introduced example of the secondary control reserve market: DER might participate at the energy market but they could never participate in the capacity market because their generation is uncertain. The authors already point out that this concept can only be implemented with improved information and communication technology. Virtual power plants are closely related to the concept of sector coupling. Sector coupling describes the integration of different forms of energy usage. This typically includes among others the power, heat and transportation sectors. A combination of DER and electric vehicles to create virtual power plants is described by Schuller et al. (2015). Technologies that enable sector coupling are power-to-gas electrolyzers, electric vehicles, vehicle-to-grid services or heat pumps, just to name a few. Sector coupling allows for electrification of energy usage that has traditionally relied on fossil energy sources such as space heating or transportation. This requires, however, a drastic increase in renewable generation capacity to ensure further decarbonization. On the other hand, it facilitates the integration of fluctuating renewable energy because the concept comes with more storage ability. It is easier to store heat energy than electric energy. Strong market penetration of electric vehicles will lead to large mobile battery storage capacities. Therefore, sector coupling will champion the concept of the virtual power plants because it finally creates flexibility in the electricity system that can then be combined with fluctuating renewable generation to make its capacity "visible." This will require much more coordination that can only be ensured through the Smart Grid. On the market side, this potential will be coordinated by aggregators who contract different resources in the market to combine them into a capacity product. These aggregators will have to combine the resources such that they can optimally position them in the different electricity markets. The optimal composition of these portfolios is described by Gärttner et al. (2018). These developments might not necessarily change the market. However, from a market perspective, it might be important to change the rules such that this flexibility can be integrated profitably. For instance, consumption of self-produced electricity might be exempt from certain fees to encourage the installation of decentralized resources and the electrification of the residential heat and transportation sector, which, in turn, increases the system's flexibility. This means that the Economic and Legal Environment needs to be adapted before such markets can become reality. This is out of the hands of the market engineer, but it sharpens our understanding of the situation. Ultimately, a shift in this foundation can create new market actors that increase the efficiency of the overall market through increased competition which leads to changing agent behaviour.

Congestion markets. Virtual power plants as described in the previous paragraph can be marketed on existing power markets such as the wholesale market or reserve power markets. Their flexibility is compensated either for consumption in cheaper time periods or for the balancing of short-term fluctuations. Another increasingly important aspect is grid congestion, both in the transmission, but more importantly, in the distribution grid. Transmission grid congestion is a growing concern in Europe (Lang et al. 2020). Another problem on the horizon that is still a rare event at the moment is distribution grid congestion. Such situations have been occurring infrequently when PV or wind energy is fed into the low voltage grid while consumption is low and the capacity is not sufficient to transmit the energy to the high voltage grid (Schermeyer et al. 2018). In the future, it is possible that similar situations might occur in the other directions when many electric vehicles try to charge at the low voltage level at the same time, or if the market penetration of heat pumps increases. This might be resolved through market-based solutions (Flath et al. 2014). Such markets might be implemented using the Smart Grid by providing short-term price signals that incentivize households to charge their electrical storage, electric vehicles or heat storage in times of local congestion. However, the design of such markets is very controversial. Proponents argue that market coordination might increase the investment in flexibility potential such as electrical storage (Huber et al. 2018). Opponents argue that such markets create gaming opportunities and will increase the costs for congestion management (Hirth et al. 2019). This discussion only shows that the solution to the problem of regional congestion is yet to be found. New markets and products are necessary which are enabled by the Smart Grid to reward regional flexibility when needed. Regional congestion or flexibility markets are ultimately another platform for virtual power plants to market their flexibility. For such markets, all components need to be defined. First, the Economic and Legal Environment needs to be adjusted to allow for such markets. However, the European Commission requires member states to implement market-based congestion management solutions (Hirth et al. 2019). Then, the transaction object needs to be defined. This could be energy in a consecutive market after the spot market clearing as discussed in (Hirth et al. 2019). However, it might also be capacity similar to the secondary reserve power market. Then, the market structure needs to be defined. We leave this as an exercise for the readers. Your objective is to achieve an efficient market outcome that mirrors a nodal pricing optimization with competitive bids by the participating agents.

Smart energy communities. An emerging concept that has attracted much attention from researchers and practitioners is that of citizen energy communities or local energy markets. In such markets, the participants trade their own generation peer-to-peer with their neighbours (Mengelkamp et al. 2017). Some countries encourage this through specific regulations. The rationale is that local balancing of supply and demand relieves the grids and that it increases the incentives for the residential population to participate in the transition to more renewable generation. The European Commission has issued a directive to support citizen energy communities which they define as "a legal entity which is based on voluntary and open participation, effectively controlled by shareholders or members who are natural persons, local authorities, inducing municipalities, or small enterprises and microenterprises. The primary purpose of a citizen's energy community is to provide environmental, economic or social community benefits for its members or the local area where it operates, rather than financial profits. A citizen's energy community can be engaged in electricity generation, distribution and supply, consumption, aggregation, storage or energy efficiency services, charging services for electric vehicles or provide other energy services to its shareholders or members." (European Union 2019). The rise of the concept of citizen energy communities is closely tied to the advent of blockchain technology as a means to allow for the creation of markets that operate without central intermediary. This concept was first introduced commercially through the Brooklyn Microgrid (Mengelkamp et al. 2018). This has sparked a variety of local energy market pilots (Weinhardt et al. 2019). Currently, the value of such designs is mostly symbolical: The trading of local energy might make the transition to renewable generation more tangible for the population and the associated regulated financial incentives lead to the consideration of decentral, green technology. In the long run, these markets might become important platforms for the coordination of demand and supply in a local area that respects constraints with the higher voltage grids. Citizen energy communities are of course highly dependent on Smart Grid technology. They can only be realized if participants can record their production and consumption in real time. From a market engineering perspective, this concept is highly interesting. As households are virtually unable to forecast their demand, it is impossible to trade ahead of time. Therefore, both control and market clearing need to happen in real time. This is a challenge for the microstructure of the market as well as for the IT infrastructure. Possible solutions are proposed in Wörner et al. (2019) or Richter et al. (2019).

4.2 Consumption Flexibility

The previous section outlines the active inclusion of new resources such as virtual power plants or markets for new resources that add flexibility to the system. This implies active bidding of participants on market places such as the wholesale markets or in peer-to-peer energy markets. Similar effects might be achieved through unilateral market signals to consumers in the form of specific tariffs. Such tariff options can have a temporal and a spatial component. Real-time tariffs signal the marginal cost of production for electricity at any given time. This might incentivize automated consumers such as heat pumps or electric vehicles to consume when cheap electricity is available. The famous case of a washing machine that starts when energy is cheaply available is, however, more unlikely to have a substantial impact: The very small cost-saving potential hardly justifies any active behavioural change. Besides real-time prices, there is a variety of pricing and regulation concepts that a regulator or utility can introduce to change residential behaviour. The most prominent ones are discussed in the following paragraphs.

Net metering. Net metering is a concept that is especially present in retail energy markets in the United States. However, a similar concept is commercially available in Germany in combination with electrical storage and labelled as the "electricity cloud". The German regulator has recently proposed a similar concept in relation to self-consumption. In essence, it means that self-generated energy can be fed into the grid at any time for the current retail price. In case of a flat tariff, this means that the grid essentially serves as a battery storage for prosumers. They can feed their excess generation into the grid at any time and their electricity meters run backwards. Then, they consume the same amount at any later time and their electricity meters run forward again. More detailed explanations are provided by Eid et al. (2014). This concept is often criticized because the cost for balancing the infeed is left for the system operators and occurring costs need to be distributed across all consumers even those without their own generation. So far, the concept has not been further developed. However, it might have the potential of serving as an incentive mechanism for the integration of fluctuating renewable generation. For instance, a household with a PV power plant could be given specific times when feed-in is calculated towards an account that can then be used at other specific times. Therefore, the load serving entity could incentivize a local feed-in management that is beneficial to the system through an efficient tariff. Such a tariff design could incentivize households to add technology that allows a shift of consumption to other times or small storage units that could shift generation at least by a little. In this case, the grid would be used as a battery but in a way that is beneficial for the overall system. From a market engineering perspective, it could be attractive to devise a platform that collects and matches supply offers and net metering tariffs between prosumers and electricity suppliers.

Peak pricing. Another tariff concept is peak pricing. Its underlying idea is to distribute system costs to those consumers who cause the need for system expansions, like grid reinforcements or generation capacity investments (in countries with capacity markets) (Burger et al. 2020). Therefore, consumer prices are much higher during system peak times. This is useful because existing electricity distribution and transmission has no marginal price. Only the need for expansions causes additional costs. Fixed charges and volumetric electricity rates, however, do not incorporate these costs. The Smart Grid could enable consumers to react to peaks quickly to avoid surcharges. This could be used in a variety of ways: Charges could be increased during peak times. On the other hand, feed-in tariffs could be lowered during times of excessive supply. Peak pricing on a local level could help to avoid local congestion and reduce the need for grid expansions. A corresponding example is provided by Flath et al. (2014) to residential EV charging. In some markets, critical peak pricing is already a common concept. Especially, if applied regionally in low voltage grids, it will support the integration of fluctuating renewables and new appliances with high energy consumption. This concept does not necessarily change the market, but it certainly increases demand flexibility. This increases the elasticity of the demand curve, and therefore, increases the efficiency of the electricity market leading to better protection against blackouts among other things (Cramton and Stoft 2005).

Time-variable tariffs. Net metering and peak pricing are special cases of time-variable tariffs that might include additional spatial components. That means that prices do not only vary with time but they might also vary depending on the location of the consumption. Time-variable tariffs include electricity prices that vary by time, signalling that electricity consumption causes different costs at different times. If expensive conventional power plants need to be ramped up to cover the last few kilowatt-hours of demand, then this is reflected in the price signal of time-variable tariffs (Burger et al. 2020). There are many forms of these tariffs. Probably the easiest form is that of a night and day tariff. Electricity is then usually cheaper at night because the overall consumption is lower and the generation occurs through less expensive power plants. These tariffs have a long history and can be realized through analogue technology using two different electricity meters that measure the consumption during the day and at night. Besides such simple time-of-use tariffs with two levels, time-variable tariffs can also include more granular price signals, e.g. with three price levels or even hourly levels. Furthermore, prices could vary across weekdays or seasons. Commonly, such tariffs are set in advance by the utility, communicated to the customer, and only updated on an annual basis. The most granular form of a time variant electricity tariff constitutes real time pricing, where price signals are calculated and communicated in real time. More elaborate time-variable tariffs become possible through Smart Grid technology as consumption is recorded in real time and because it allows the communication of signals from the grid to the consumer who can use automated appliances that react to the signals as described by Dauer et al. (2016). However, as electricity is a basic necessity for all households, it should be considered that such changes in tariff design can have adverse socio-economic consequences, for example, for low-income households. A very good discussion on different tariff designs and their socio-economic effects is given by Burger et al. (2020). The described forms of time-variable tariffs reward flexibility directly if it is executed. However, the communication of available flexibility might also be of value to system operators and utilities. Such flexibility can be actively used to balance suddenly increasing or decreasing generation as in case of a passing cloud. Tariffs can be used to incentivize the communication of such flexibility. One such tariff design is deadline differentiated pricing which is described by Salah and Flath (2016), among others. This design rewards consumers if they provide a later deadline for a stated consumption goal. It allows operators to react to the actual generation and schedule different demands more efficiently. One possible use case is a parking garage with solar panels. Electric vehicles can then provide their desired state of charge at any given time in the future. If the parking garage operator is given longer deadlines, it is easier for her to ensure full satisfaction of demand through self-consumption of the generated solar power. This again requires communication between the consumer and the generator and an exact record of consumption which can all be facilitated through the Smart Grid. A market engineer can design platforms that serve as intermediaries between suppliers and consumers. Consumers can enter their flexibility potential which is then offered to suppliers that can react with tailored offers of time-variable tariffs that range from real time pricing to simple flat-rate tariffs. Here, the IT infrastructure that translates flexibility into tariff offers, is the most complex market component (see vom Scheidt et al. (2019) for an initial approach).

5 Market Evolution

In the previous section, we introduce ways for the Smart Grid to be used to improve the operation of electricity markets. Real-time signals can be used to incentivize the activation of additional resources, thus increasing competition or they can cause appliances to react to price signals, thereby increasing the demand elasticity. Both measures increase the efficiency of the market and thus improve the market outcome. In this section, we discuss the process of adaption that occurs in the medium and long-term through market incentives. These adaptions then have an effect on the short-term operation of the electricity system and increase its efficiency through the described channels. Such adaptions in terms of investment are supported by the Smart Grid through the provision of time series data (compare chapter "Smart Grid Analytics") or through the possibility of coordination that makes the investment in shared resources possible. The increasing information availability can also allow retailers to develop new business models (compare chapter "Business Model Design") to increase customer satisfaction. Such models often include coordination mechanisms. Thus, the role of the market engineer is to develop market mechanisms that are provided as part of holistic solutions that increase the efficiency of the energy market. Three applications of market mechanisms are described in the following. This is by no means an exhaustive list and readers are encouraged to find more applications and share them with us (or get rich by themselves).

Hardware investment. Private households and industrial consumers can benefit from the investment in generation and storage hardware. While industrial consumers are aware of their consumption, private households usually undertake such investments either because feed-in tariffs allow for a broad estimation of amortization times or because of a combination of standard load profiles and a broad gut feeling that such investments will pay off. The fact that this is already possible is underlined by several startups in this field such as sonnen and their battery solutions in Germany. The Smart Grid and the resulting information availability allow private consumers to estimate the benefits of additional power hardware more precisely. This can lead to more accurate recommendations for private households which might include alternative tariffs or the marketing of such hardware that is described in the previous section. Such recommendations can be given on an individual level or platforms can enable the connection of several households in a neighbourhood. Local PV generation or storage solutions might be more attractive if the corresponding consumption profiles fit well together. Such a platform is described by Golla et al. (2020). The capacity of these investments might later be traded at other markets by the provider, the utility or the customers themselves. For instance, sonnen already markets a virtual connection of their residential battery storage systems at the primary control market (Angenendt et al. 2020). Various developments are possible in this field which is very attractive in terms of future business models. Providers could also offer the hardware for reduced rates in turn for the permission to use consumption data for other purposes such as advertising. This field will, therefore, become a very diverse market without products that can easily be defined as the transaction object in the market engineering framework. This is further discussed in chapter "Case Studies in the Smart Grid Sector".

Community storage. One specific use case of hardware investment is the shared use of resources by multiple households (Golla et al. 2020). This is especially attractive for storage as it can increase the used capacity if combined with residential PV capacity either in times of low PV generation or if residents are unable to perform one full cycle per day with their storage capacity due to their low consumption. Sharing storage capacity can incentivize the investment in more capacity, which in turn supports the energy transition through local balancing of supply and demand and thereby reduces the transmission and distribution requirements of the electricity grid. Some initial studies have been performed, for instance, by Barbour et al. (2018). The exact product and market definition of community storage capacity is still subject to further research. This is an interesting field for market engineers: The market has two perspectives, that of consuming from the storage as well as that of charging the storage. The transaction object could be storage capacity over time, but it could also be simply electricity that is consumed from the storage. The design needs to balance the interests of storage and PV owners, PV owners and simple consumers. Everyone, who benefits the system somehow needs to benefit from the participation in the market to ensure individual rationality. However, to which extent individuals benefit is subject to further analysis and even depends on the subjective understanding of fairness of the participants. The necessary market design includes all aspects of the market engineering framework and it is an interesting task to come up with different designs.

Tariff platforms. As in the previous section, we can differentiate the market evolution between adapting hardware and reacting to price signals. For market evolution, the latter means a change of the residential electricity tariff. While current electricity tariffs are usually rather simple and do not reflect spatial and temporal costs of electricity, this might change in the future. However, it is difficult to choose an elaborate electricity tariff without in-depth knowledge of the personal future consumption patterns, especially if no smart meter is installed. Smart Grid technology allows retailers to recommend tariffs more precisely. Knowledge about appliances in the household can be used to quantify demand response potential. The exact appliance endowment can be characterized using non-intrusive load monitoring, a data science technique that uses exact load profiles to identify individual appliances (Zoha et al. 2012). Furthermore, personal load curves can be used to assess the consumption profile with regards to system peaks and real-time prices. This way, platforms can recommend changing electricity tariffs for individual households. This can go hand-in-hand with recommendations for new appliances such as energy management systems. These can react flexibly to market signals and charge electric vehicles or a heat storage in times beneficial for the user. A market engineer can design complex markets, where the installation of appliances in combination with specific tariffs leads to benefits for the consumer. This might evolve into a Smart Grid market platform, another step further from mere tariff recommendation. Such recommendations have already been studied in vom Scheidt et al. (2019). The authors assess different electricity tariffs and the possibility to predict the optimal tariff based on consumption data of only one month. However, they do not use sophisticated tools, but simply assess what would happen if the best tariff for that month was to be adopted. In markets with a large variety of tariffs and tariff-hardware bundles, it will be important to provide a product that supports customers with their decision or even takes away their risk in return for some concessions like giving up the control over their heat pump. Therefore, the design of the transaction object will play an important role when it comes to tariff platforms.

6 Summary

In this chapter, we describe market engineering as an important task for the energy system. We introduce the market engineering framework as a tool that can be used to describe existing markets and to design and engineer new markets for innovative transaction objects. Furthermore, we describe a set of important market characteristics that need to be taken into consideration when markets are designed. These fundamentals are important to understand how the Smart Grid can support changes in market designs and how it can be leveraged to introduce new markets that increase the efficiency of the electricity system. We describe neuralgic points that influence the market result and are affected by the Smart Grid. Finally, we provide examples of how the Smart Grid changes the electricity market and describe the role of the market engineer within these changes. The provided examples can be classified along the electricity value chain and the market aspect influenced by the Smart Grid. The value chain is classified into generation, transmission and distribution and retail. The influence of the Smart Grid is divided into increasing competition, fostering consumption flexibility (which can be understood as effects on market operation) and providing market information which mainly impacts the market evolution. We provide examples in each category, which are in no way exhaustive. Readers are encouraged to add to these lists and to send us suggestions to enter an academic discussion. The Smart Grid adds functionality, especially in the areas of appliance control, communication of market signals and through increased data availability that can support new products. There is a variety of opportunities that can be leveraged and the coming years will show which areas evolve the quickest.

7 Exercises

Competitors. Imagine the following setup of two prosumers with PV panels and some additional non-flexible demand. There is also an electric vehicle with a capacity of 6 kWh that is fully charged at the beginning of this period and which can be charged and discharged fully within one timestep. It has a need for 6 kWh by the end of the period. There is a CHP plant that can generate electricity at 10 cents per kWh. Self-

Table 2 Local energy system

Timestep	1	2	3	4	5	6
PV1	0	0	1	4	1	0
Consumption1	1	2	3	1	1	2
PV2	0	0	1	8	2	0
Consumption2	0	1	2	2	1	1
Consumption rest	0	1	4	1	3	1

consumed PV electricity has marginal costs of zero and no charges. It is rewarded with 5 cents per kWh if fed into the grid through a feed-in tariff. This is also paid if the renewable generation needs to be curtailed. Detailed information on the generation and consumption in the electricity system is provided in Table 2.

1. PV prosumer 1 combines his PV panel with the electric vehicle through vehicle-to-grid technology to form a virtual power plant. What is her maximum revenue during the period given the consumption she can replace?
2. Assume that the transmission capacity of the community to the higher voltage level is 5. Assume that the electric vehicle acts at a congestion market. What is its maximal revenue of the electric vehicle without increasing the system costs?
3. Assume that an energy community can act as one entity and self-consume locally generated PV. What is the global cost benefit of the community compared to the case of individually acting agents? How would you distribute this gain?

Flexible Demand

1. Assume the same prosumer profiles as in Exercise 1. What is the system benefit in terms of reduced need for curtailment payments if net metering was performed such that 8 cents per kWh were paid as feed-in tariff in the first and last period and no feed-in tariff at all in period three and four?
2. Assume that during the daily peak each kWh costs an extra 5 cents. Assume further that consumer 1 and consumer 2 can shift one kWh freely during the period and that the remaining customers can shift 2 kWh freely. Ignore the electric vehicle. Construct the final load profile for the entire period.
3. Assuming that each consumer can shift two units of their load into the next timestep and that every consumer can shed one unit of load over the entire period. The real time prices per timestep are given as p_t. Formulate the optimization problem of shifting load optimally for each consumer given that the electric vehicle belongs to consumer 1.

Market Information

Pecan Street gibt es nicht mehr kostenlos

1. Download the SCiBER power consumption data set (https://im.iism.kit.edu/sciber.php) and choose one consumer at will. Then use the renewables.ninja tool to generate a PV generation curve (https://www.renewables.ninja/). Use one month of consumption and PV generation data to calculate the benefit of a PV panel and a 5 kWh storage unit that costs 8,000 \$. Then evaluate the payback of the investment over the remaining period. How long is the amortization period? Assume grid electricity costs of 20 cents/kWh.
2. Use the introduced market engineering framework to design a market for a community storage unit. You can freely design the transaction object.
3. Using the SCiBER data, calculate the optimal tariff for each household assuming the tariff options from (vom Scheidt et al., 2019) with each month of the data. Then, compare the results with the global optimum over the entire data. Construct a confusion matrix that shows which tariff was and which should have been recommended.

References

Alaywan Z, Wu T, Papalexopoulos AD (2004) Transitioning the California market from a zonal to a nodal framework: an operational perspective. In: IEEE PES power systems conference and exposition, 2004. IEEE, pp 862–867

Angenendt G, Zurmühlen S, Figgener J, Kairies K-P, Sauer DU (2020) Providing frequency control reserve with photovoltaic battery energy storage systems and power-to-heat coupling. Energy 116923

Barbour E, Parra D, Awwad Z, González MC (2018) Community energy storage: a smart choice for the smart grid? Appl Energy 212:489–497

Bompard E, Ma Y, Napoli R, Abrate G (2007) The demand elasticity impacts on the strategic bidding behavior of the electricity producers. IEEE Trans Power Syst 22(1):188–197

Burger SP, Knittel CR, Pérez-Arriaga IJ, Schneider I, Vom Scheidt F (2020) The efficiency and distributional effects of alternative residential electricity rate designs. Energy J *41*(1)

Cramton P, Ockenfels A (2016) Economics and design of capacity markets for the power sector. In: Interdisziplinäre Aspekte der Energiewirtschaft. Springer, pp 191–212

Cramton P, Stoft S (2005) A capacity market that makes sense. Electr J 18(7):43–54

Dauer D (2016) Market-based allocation of local flexibility in smart grids: a mechanism design approach. PhD thesis, Karlsruher Institut für Technologie (KIT)

Dauer, D, Vom Scheidt F, Weinhardt C (2016) Towards smart distribution grids: a structured market engineering review. In: 6th Karlsruhe Service Summit, Karlsruhe, Germany. Grin Publishing, München, Germany

Eid C, Guillén JR, Marín PF, Hakvoort R (2014) The economic effect of electricity net-metering with solar PV: Consequences for network cost recovery, cross subsidies and policy objectives. Energy Policy 75:244–254

Union European (2019) Directive (EU) 2019/944 of the European parliament and of the council of 5 June 2019 on common rules for the internal market for electricity and amending directive 2012/27/eu. OJ L 158:125–199

Flath CM, Ilg JP, Gottwalt S, Schmeck H, Weinhardt C (2014) Improving electric vehicle charging coordination through area pricing. Transp Sci 48(4):619–634

Gärttner J, Flath CM, Weinhardt C (2018) Portfolio and contract design for demand response resources. Eur J Oper Res 266(1):340–353

Golla A, Henni S, Staudt P (2020) Scaling the concept of citizen energy communities through a platform-based decision support system. In: Twenty-Eigth European conference on information systems (ECIS2020), Marrakesch, Marocco

Hirth L, Glismann S (2018) Congestion management: from physics to regulatory instruments

Hirth L, Schlecht I et al (2019) Redispatch markets in zonal electricity markets: Inc-dec gaming as a consequence of inconsistent power market design (not market power)

Huber J, Köppl S, Klempp N, Schutz M, Heilmann E (2018) Engineering smart market platforms for market based congestion management. In: Proceedings of the ninth international conference on future energy systems, pp 544–549

Kyritsis E, Andersson J, Serletis A (2017) Electricity prices, large-scale renewable integration, and policy implications. Energy Policy 101:550–560

Lang LM, Dallinger B, Lettner G (2020) The meaning of flow-based market coupling on redispatch measures in Austria. Energy Policy 136:111061

Mas-Colell A, Whinston MD, Green JR et al (1995) Microeconomic theory, vol 1. Oxford University Press, New York

Mengelkamp E, Gärttner J, Rock K, Kessler S, Orsini L, Weinhardt C (2018) Designing microgrid energy markets: a case study: the Brooklyn microgrid. Appl Energy 210:870–880

Mengelkamp E, Staudt P, Garttner J, Weinhardt C (2017) Trading on local energy markets: a comparison of market designs and bidding strategies. In: 2017 14th international conference on the European energy market (EEM). IEEE, pp 1–6

Pudjianto D, Ramsay C, Strbac G (2007) Virtual power plant and system integration of distributed energy resources. IET Renew Power Gen 1(1):10–16
Richter B, Mengelkamp E, Weinhardt C (2019) Vote for your energy: a market mechanism for local energy markets based on the consumers' preferences. In: 2019 16th international conference on the European energy market (EEM). IEEE, pp 1–6
Salah F, Flath CM (2016) Deadline differentiated pricing in practice: marketing EV charging in car parks. Comput Sci-Res Dev 31(1–2):33–40
Schermeyer H, Vergara C, Fichtner W (2018) Renewable energy curtailment: a case study on today's and tomorrow's congestion management. Energy Policy 112:427–436
Schuller A, Flath CM, Gottwalt S (2015) Quantifying load flexibility of electric vehicles for renewable energy integration. Appl Energy 151:335–344
Staudt P (2019) Transmission congestion management in electricity grids—designing markets and mechanisms. PhD thesis, Karlsruher Institut für Technologie (KIT)
Vom Scheidt F, Staudt P, Weinhardt C (2019) Assessing the economics of residential electricity tariff selection. In: 2019 international conference on smart energy systems and technologies (SEST). IEEE, pp 1–6
Walling R, Saint R, Dugan RC, Burke J, Kojovic LA (2008) Summary of distributed resources impact on power delivery systems. IEEE Trans Power Deliv 23(3):1636–1644
Weinhardt C, Holtmann C, Neumann D (2003) Market-engineering. Wirtschaftsinformatik 45(6):635–640
Weinhardt C, Holtmann C, Neumann D (2003) Market-engineering. WIRTSCHAFTSINFORMATIK 45(6):635–640
Weinhardt C, Mengelkamp E, Cramer W, Hambridge S, Hobert A, Kremers E, Otter W, Pinson P, Tiefenbeck V, Zade M (2019). How far along are local energy markets in the dach+ region? a comparative market engineering approach. In: Proceedings of the tenth ACM international conference on future energy systems, pp 544–549
Woo C-K, Lloyd D, Tishler A (2003) Electricity market reform failures: UK, Norway, Alberta and California. Energy Policy 31(11):1103–1115
Wörner A, Ableitner L, Meeuw A, Wortmann F, Tiefenbeck V (2019) Peer-to-peer energy trading in the real world: market design and evaluation of the user value proposition
Wurman PR (1999) Market structure and multidimensional auction design for computational economies. Citeseer
Zoha A, Gluhak A, Imran MA, Rajasegarar S (2012) Non-intrusive load monitoring approaches for disaggregated energy sensing: a survey. Sensors 12(12):16838–16866

Regulatory and Institutional Aspects of Smart Grids

Check for updates

Gert Brunekreeft, Marius Buchmann, and Anna Pechan

Learning Objectives

- To be able to describe different forms of unbundling of generation from grids and to understand its renewed relevance in the context of smart grids and data management
- To be able to describe different approaches to network regulation including their benefits and drawbacks and to get an insight into current developments
- To be able to describe principles and predominant forms of network pricing and to get a grasp on how smart meters may influence it
- To get an insight into current and potential forms of (market-based) congestion management in smart grids.

1 Introduction on Electricity Market Structure

For an economist, the power system value chain consists of four main stages: generation, transmission, distribution and retail. Generation is the production and retail is the sale of electricity to end-users; sometimes, wholesale trade is considered another separate stage in the value chain. These are potentially competitive stages, where

G. Brunekreeft (✉) · M. Buchmann · A. Pechan
Bremen Energy Research, Jacobs University, Campus Ring 1, 28759 Bremen, Germany
e-mail: g.brunekreeft@jacobs-university.de

M. Buchmann
e-mail: m.buchmann@jacobs-university.de

A. Pechan
e-mail: a.pechan@jacobs-university.de

C. Dinther et al. (eds.), *Smart Grid Economics and Management*,
Lecture Notes in Energy 51, https://doi.org/10.1007/978-3-030-84286-4_5

commercial companies can unfold market activities. In between, we find the networks; first the high-voltage transmission network and second the medium- and low-distribution networks. These are natural monopolies. A natural monopoly exists when it is cheaper to provide a certain industry output by a single firm than by several competitive firms (Baumol 1977). This is typically the case in industries with high fixed costs, which do not vary much with the output, and low marginal production costs. Applied to the energy supply industry, this means that connecting and serving all users in a certain area by one monopolistic electricity network operator incurs lower total costs as if separate competitive network operators would provide the service for subgroups of these users. There may be many such network companies, but they are all monopolies in their own services area. Figure 1 depicts these stages of the electricity value chain.

The vertical value chain sets the academic fields of regulation, competition policy, industrial organization and market design and sets the background for this chapter. The commercial companies (here generation and retail) are actors in competitive markets, for which they need access to the monopoly networks and operating systems. This network access needs to be non-discriminatory among all commercial firms. The problem arises when one of these commercial firms may be the affiliate of the network company; we call this vertical integration. The problem is that the vertically integrated company may have incentives to discriminate third parties on the commercial markets to the advantage of its own commercial affiliate: this is vertical leverage of market power (i.e., leverage from the network to the commercial stage). To address this problem, regulatory authorities have implemented rules for network unbundling: the main aim is to either curb the potential for leverage of market power (in case of vertical integration) or, one step further, to take away the incentives for this leverage by ownership unbundling (vertical separation). There are many variations of unbundling concepts, all with pros and cons; there is no single best solution. So far,

Fig. 1 The electrical energy value chain

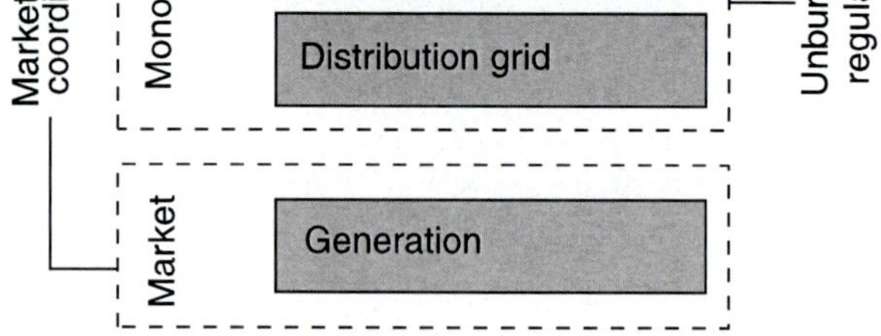

the discussion applied mostly to the transmission level. Following the developments to smart distribution grids, we observe that a similar debate starts at the distribution level. We discuss this in Sect. 2.2.

The networks are natural monopolies. They are subject to revenue or profit regulation for two main reasons. First, to secure non-discriminatory third-party access for all players on the commercial stages. Second, to protect the end-users against monopoly power. A regulation mechanism should set incentives for the regulated companies for efficient operation on the one hand and facilitate investment on the other hand. Several decades of practical experiences suggest that this is an uneasy combination of goals. With cost-based and price-based approaches, we find two basic regulatory models; these are extremes and, in practice, we find many hybrid forms. Interestingly, arguably driven by the energy transition, following the new trend that network operators are facing many new task and roles, a new type of regulation is currently developing: output-oriented regulation. We discuss this in Sect. 3.4.

Network users, be it companies or end-users, pay for network usage for different reasons. First, the networks need to be financed. Second, network charging sets signals for the network users to optimize short-term use and long-term development of the network. For a long time, the networks were mostly uncongested. This has changed: network congestion is rather the rule than the exemption. Network congestion management now tries to set signals for optimal short-term network use and optimal long-term investment. Here is where smart network pricing and contracting and smart markets step in. A topical development is decentralized congestion management: network congestion increasingly needs to be managed with congestion-relieving demand and supply at the distribution level; we call this flexibility. The current debate is how to organize decentralized flexibility markets to relieve network congestion. We discuss this in Sect. 4.2.

2 Governance Models for Smart Grids

Within the last decades, the electricity supply chain (compare Sect. 1) was primarily shaped by the liberalization process (Joskow 1996). Before liberalization, a hierarchical and integrated system existed in the electricity sector. Utilities were active in all stages of the supply chain with one and the same vertically integrated company. However, in their seminal work in 1983, Joskow and Schmalensee (1983b) point out that the introduction of competition in generation could increase the overall efficiency of the electricity sector, which started a process known as structural reform. The extent of this structural reform differs between countries; we focus our analysis on the structural reforms in Europe and add experiences from other regions where appropriate.

2.1 *The Structural Reform of the Electricity Supply Chain in Europe—Before Smart Grids*

Today, network unbundling is the norm in Europe: unbundling describes the separation of the natural monopolies (i.e., electricity networks) from the competitive parts of the supply chain (namely, generation and retail). In its most extreme form, this means that the unbundled networks are owned and operated by (independent) companies that are not active in generation or retail. Or vice versa, generation or retail companies do not own the networks. The liberalization process in the European Union on the transmission level involved three steps, starting with the First Electricity Directive of 1996 (European Commission 1996), which was followed by the Second Electricity Directive in 2003 (European Commission 2003) and the Third Directive in 2009 (European Commission 2009). The European Commission pursued four goals by liberalizing the electricity market (for details see Meyer 2012): The main goal of the liberalization process in the EU was to establish a single European electricity market. Second, liberalization was established to secure third-party access to the markets in generation, trade and retail. Third, third-party access to the network infrastructure was regulated to prevent discriminatory behaviour by network owners against other generation and retail companies (see Sect. 4). Fourth, final customers should be allowed to choose their electricity supplier, called retail competition.

The results of the structural reforms in Europe differ between the network levels, the transmission and distribution networks.

The Institutional Framework on the Transmission Grid Level
The current institutional framework in the EU is based on the 3rd legislative package (European Commission 2009). Thereby, the Commission introduced three different options for unbundling on the transmission level, i.e.

- ownership unbundling,
- (deep) Independent system operator (ISO),
- Independent transmission operator (ITO)

Figure 2 provides an overview of the different governance models, which include the three options above. We will introduce the other governance models later in this section. Full ownership unbundling prohibits joint ownership of network and generation or retail assets within one firm.

The ITO model allows companies to retain both network ownership and management, but it puts strong limitations on cross involvement of employees to assure independence of the network. The ITO scheme is in effect a stronger form of legal unbundling. Legal unbundling was introduced in the Second Electricity Directive in 2003 (European Commission 2003) and requires that the network operator is independent at least in terms of its legal form, organization and decision-making from other activities not relating to transmission (i.e., generation and retail). This includes unbundling of accounts, operations and information. The idea behind this is to ensure that no relevant information is exchanged between the network and other parts of the supply chain within one utility. One can think of legal unbundling as "firewalls"

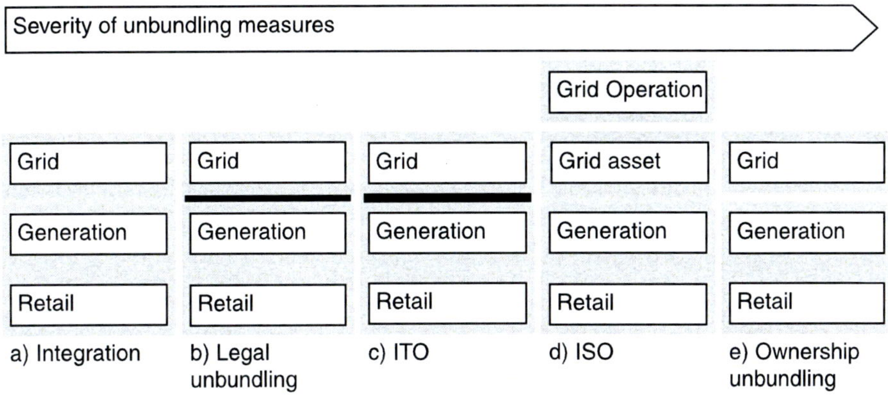

Fig. 2 The different unbundling schemes in the electricity sector and their degree of vertical separation. **a** Applies to all DSOs in Europe that serve less than 100,000 customers. **b** All DSOs with more than 100,000 customers are legally unbundled. **c** ITO is a stronger form of legal unbundling. **d** Ownership unbundling describes the full separation of the networks from all other parts of the supply chain. **e** ISO separates network ownership (still integrated) from network operation (separated)

or "Chinese walls" which prohibit such a flow of information within one integrated company (Brunekreeft and Keller 2001). Still, a legally unbundled network operator can be part of a holding company that owns generation and retail subsidiaries as well.

Conceptually, we can think of a transmission system operator (TSO) to consist of the owner of the transmission assets (TO) and the system operator (SO). These are usually one and the same company, but that need not be. The unbundling option ISO requires that an independent entity takes over operational activities (system operation) in the network, separate from transmission asset ownership. With an ISO, the network ownership can stay with the integrated firm, which also owns generation assets. The primary task of the ISO is scheduling and dispatching generation and load and the allocation of scarce transmission capacity. In addition, the ISO cooperates with the transmission owners and other stakeholders to coordinate maintenance schedules and plans new transmission investments together with the transmission owner. As the ISO has no direct interest in the financial performance of the owners of any of the assets that comprise or utilize the transmission network, it can be expected to be neutral (Balmert and Brunekreeft 2010).

In Europe, transmission system operator (TSO)s are either ITOs or ownership unbundled. With very few exceptions (e.g., the UK), the ISO model is not applied in Europe. Rather, the ISO model is the standard in the US, where the transmission networks as operated by independent ISOs, so-called Regional Transmission Organisation, but the network assets are owned by separated utilities that own both, the transmission and distribution networks, in addition to generation and retail.

The Institutional Environment on the Distribution Grid Level

In Europe, as well as in the US, the institutional environment on the distribution grid level is different compared to the previously described framework on the transmission level. In the EU, distribution networks are currently subject to legal unbundling as a minimum requirement (European Commission 2009). However, legal unbundling is only applied for those DSOs that have more than 100,000 customers. DSOs with fewer customers do not have to unbundle and can remain an integrated part of a utility. This exception is known as the de-minimis rule (specified in European Commission 2009, Art.26). Out of the roughly 880 DSOs in Germany, for instance, only a small share (about 150) has such a large customer base, which, in turn, means that roughly 80% of all DSOs are still part of integrated utilities (European Commission 2011). It needs to be noted that the legally unbundled DSOs, which are not subject to the de-minimis rule, own large parts of the overall network in most member states of the EU. Typically, these larger DSOs own roughly 95% of the national networks (even though their number is quite low), exceptions are Denmark (small DSOs own 43% of networks) or Austria (12% of the networks in the hands of small DSOs). In the US, on the other hand, DSOs are fully integrated with generation and retail (in most states).

Notwithstanding the advantages of promoting more competition, unbundling does have drawbacks and was met with criticism. Joskow and Schmalensee (1983a) already stressed that the unbundling of the networks will require complex coordination mechanisms (i.e., contractual relations) to substitute the previously internal planning processes of integrated utilities. At the heart of the discussion about the coordination mechanism are transaction costs, which are the costs related to establish a contractual relation, e.g., the costs to identify partners, define the contract, etc. (see Coase 1937; Williamson 1979, for details on transaction cost theory). Basically, information exchange in integrated utilities results in lower transaction costs as does the information exchange between separated entities, as long as the market has not established efficient coordination mechanisms (i.e., low transaction costs). In the electricity sector, coordination becomes especially relevant at the intersection between the networks and the electricity generation market. Here, a coordination problem evolves due to missing information exchange between the generation companies and network operators (see Brunekreeft 2015, for details). The result of the missing information exchange and the weak coordination mechanisms is an increase in costs and a decrease in efficiency, especially on the distribution grid level. Although the coordination problem has originally been a consequence of the liberalization process, its relevance increases with the diffusion of distributed generation based on renewable energy sources (RES) as the number of parties that need to be coordinated within the network increases.

With smart grids, from an institutional perspective, the challenge arises to find the right balance between two general principles: On the one hand, the institutional framework should facilitate a level-playing field, which means that all market parties, be it an incumbent energy utility that operates conventional power plants or a new market entrant that makes use of smart meter data to develop new energy services, have the same access to the electricity infrastructure. This implies that all relevant information available to the network operators, as long as this information has commercial value, is available to all market parties in a non-discriminatory way. Data on the flexibility need by DSOs is one example of such information which gains significant commercial value in smart grids. Hence, the EU Commission now requires each DSO to publish network development plans that provide information on the mid- and long-term need for flexibility (European Commission 2019). While access to networks is not a new regulatory task, access to smart meter data, for example, constitutes a new requirement for a level-playing field in smart grids. To put it differently: With smart grids, the requirement to facilitate a level-playing field extends beyond the traditional access to the electricity networks, e.g., into the digital realm as well. For example, DSOs in Europe and in the US are responsible for the smart metering infrastructure, and hence the level-playing field extends to equal access to smart meter data infrastructure as well. On the other hand, the requirement for coordination increases with smart grids as well. As described above, coordination here refers to the information exchange between the network operators and the network users. With decentralization and digitalization, the number of network users that become part of this coordination process increases. For example, distributed generation results in hundreds of thousands of independent generators, while the digital connection of electricity devices (from smart assistants to connected heat pumps, etc.) is on the rise as well. As a result, the coordination process in the electricity sector now involves a multiple of actors compared with the coordination process ten years ago.

In general, coordination within an integrated company has two advantages: First, it is relatively uncomplicated, since the different departments can just talk to each other. Second, the incentives are aligned between the different departments, since they all work for the same overall profit. Imagine an integrated utility that owns generators and networks. To coordinate the generators with the networks, e.g., to reduce the overall costs of the integration of an additional generator into an existing grid, these two departments only need to meet and exchange information in an internal meeting. Since both departments want their company's profit to increase, the incentive to cooperate will be strong. The transaction costs of this process are rather low since there is no need for a long search, etc. However, now imagine that the network operator and the generator are two independent companies, in this case, coordination cannot be facilitated by internal communication processes and the incentives are not aligned, since the separated companies focus on their own and not their combined profit. Hence, market signals are required to coordinate the investments of these two companies. If these market-based coordination mechanisms are not efficient and do not align the incentives of the involved companies, the transaction costs for coordination will be higher than in the integrated case, which, in general, is an argument for integration. However, with integration, it is more difficult to secure competition. The regulator will have a hard time to secure that the integrated network operator does not favour its integrated generator, e.g., with respect to connection charges. Hence, there is a trade-off between coordination on the one hand, and facilitating a level-playing field on the other hand. With smart grids, both criteria gain significant importance on the distribution grid level, which now drives a new debate about the level of integration/unbundling on the distribution grid level to facilitate the development of smart grids. In other words, the question is raised (and not yet answered) whether the trend towards smart grids requires an adaptation of the unbundling regime of DSOs in Europe.

2.2 The Governance Options for Smart Distribution Grids: What Changes in Smart Grids?

With smart grids, the DSOs tasks can extend beyond the traditional development, maintenance and expansion of electricity networks. For example, DSOs could take care of smart meter data management or the operation of multi-use battery storage. With these new tasks, the search for the efficient balance between facilitating a level-playing field and coordination becomes more complex. Hence, regulators in Europe are discussing whether further unbundling is deemed necessary and if so, how these unbundling models should look like. In general, the different unbundling regimes for the transmission level presented in Fig. 2 can be adapted to the distribution grid level as well.

However, the alternatives to the current legal unbundlıng of distribution grid operators all come with certain challenges. For example, ownership unbundling, which is the strongest form of unbundling, has not yet proven to be an efficient solution. De Nooij and Baarsma (2009) provide a cost-benefit analysis of ownership unbundling in electricity distribution networks, based on experiences from the Netherlands. They conclude that the ownership unbundling comes with potentially very high one-off and structural costs (e.g., for implementation of ownership unbundling rules), while current literature suggests that the related benefits might not exceed these costs. To put this differently, though ownership unbundling might be the best way to secure a level-playing field for smart grids, the potential increase in transaction costs to fully unbundle the DSOs and for coordination might exceed these benefits (Nillesen et al. 2019). Hence, from today's perspective, ownership unbundling might not be the ideal solution at hand to facilitate smart grids.

As a (less strict) alternative, the Independent distribution system operators (IDSO) model is currently discussed in the US in the context of smart grids (Burger et al. 2018). The IDSO depicts an application of the ISO model (as introduced above) to the distribution grid level. With the IDSO concept, asset ownership is separated from system operation. The asset of the network can be owned by an integrated company. System operation is delegated to an independent operator: the IDSO. Independent here means that the IDSO is not owned by or affiliated with market parties from retail, generation or other market parties like aggregators (Friedrichsen 2015). Though the concept of IDSOs is discussed more frequently with increasing decentralization, Burger et al. (2018) point out that the separation of asset ownership and operation would probably result in a lower system efficiency compared to an integrated solution. These inefficiencies are due to several key challenges that are associated with the ISO model in general, and which are relevant to the IDSO concept as well. Pollitt

(2012) summarizes these key challenges for the ISO model. Here, we point out two weaknesses of the ISO model described by Pollitt (2012) that would be relevant for the IDSO model as well:

- Complex information exchange and potential duplication of tasks: the system operator and the asset owner both need to have a highly complex system of information exchange. With smart grids, the complexity of information exchange (data exchange) will increase significantly, which potentially will raise the duplication of tasks as well.
- Costly dispute resolution procedures: if operation and asset ownership are separated, the risk allocation process can reach very complex levels. In the case of congestion management (see Sect. 4), the question about liabilities becomes very important, since the costs for the different measures (e.g., local congestion management vs. curtailment vs. investment) might differ significantly and disputes between asset owner and operator might evolve about the efficient allocation of costs.

Due to these weaknesses, the IDSO governance model does not seem to be a good alternative to legal unbundling on the distribution grid in Europe either.

This leaves us with the last governance model, the Independent distribution operator (IDO). The IDO could be considered as a governance option for smart grids with stronger unbundling of DSOs—corresponding to the ITO at the TSO level. An IDO is a stronger form of legal unbundling than it is currently applied in the EU on the distribution grid level. Although the network operator is still owned by an integrated company in this approach, it is an independent division with its own corporate identity, resources and management. The use of services from the integrated company is prohibited. The aim of these additional firewalls between the network operator and the other parts of the utility is to ensure independence from management and network investment decisions. According to the European Commission, an ITO at the transmission system level is a well-functioning alternative to ownership unbundling (CEC 2014). Furthermore, the expected additional benefits to switch from the ITO model to ownership unbundling are considered to be small (Brunekreeft et al. 2014). Together with the positive evaluation of the ITO model by the EU Commission, these insights serve as a first indicator that the introduction of an IDO to facilitate the development of smart grids in Europe might become a viable governance solution.

2.3 Case: Governance of Smart Meter Data Management in Europe

So far, the traditional analogous electricity meters were installed, operated and maintained by the distribution network operators. The DSOs thereby served as an intermediary between the network users (e.g., the consumers) and the retailers. Due to this experience, they became responsible for installing and maintaining smart meters as well, at least in most European countries (exceptions are the UK and, at least theoretically, Germany where the metering market was liberalized) and in the US. With smart meters, the commercial value of data access on the distribution grid level increases, since smart meters collect and distribute data on local energy consumption, which might provide a basis for many new business models. Due to this increased commercial value associated with smart metering, the future role of DSOs in this context is under discussion. Different European countries are in the process to establish data management systems for smart meter data. Most of these data management systems shall provide a framework to exchange data from smart metering for billing, switching processes and new tariff designs. In different European member states, the so-called data hubs are introduced to address two primary issues:

1. secure equal access to data from smart metering
2. increase efficiency in the communication between market parties, especially between network operators and retailers for billing and switching purposes

Two prominent examples for smart meter data hubs are located in Belgium and Norway.[1]

Belgium: Central market system (CMS)
Since 2018, a centralized data hub facilitates the data exchange between market parties in Belgium. This CMS is operated and financed by a company called ATRIAS, which is run by the distribution system operators. The CMS connects the databases of the network operators (who collect the data from the smart meters) with the relevant and eligible market parties. Thereby, ATRIAS has a focus on the data exchange between the DSOs and retail businesses. Other parties, like the transmission system operators and third-party service providers, shall get access to the data as well.

Norway: ElHub (Electricity Hub)
Norway has a similar plan as Belgium to manage the data exchange from smart metering. ElHub facilitates the data exchange between market parties in Norway and is operated by the national TSO. Smart metering data is collected via the DSOs and stored in the ElHub together with consumer data from the retailers. ElHub aims at standardization of data access to smart meter data for all eligible parties. In the beginning, ElHub will provide hourly values for smart metering, but might increase this up to 15-minute values. The customers are in full control of their data, which they can access via an online tool, and thereby manage third-party access to their data sets.

Different governance approaches can be applied to govern these data hubs. Here, we separate the potential approaches into two governance categories:

- *Regulated and market-based approaches* either integrate data management systems with the existing monopolies in the electricity sector, i.e., the networks, or are established as institutional monopolies granted by the government (with one entity being responsible for the national smart meter data management).

[1] Further details can be found in CEER (2016).

- *Market-based approaches* separate the smart meter data hubs from network operation. In the market-based model, any party other than the operators of the monopolistic bottlenecks could be responsible for this task. This includes incumbents from the electricity sector, as well as third parties (e.g., from the telecommunication sector), which are not yet active in the energy business. These competing data hubs would either be subject to competition law or a regulation scheme independent from the monopolistic bottlenecks of the network industries.

The decision about the specific institutional design of smart meter data management needs to take into account several factors. For example, from an institutional perspective, the chosen governance approach shall ensure non-discrimination and a level-playing field.

3 Network Regulation

In Sect. 1, it has been explained that the value chain of the electricity supply industry contains market and network stages. The networks, transmission and distribution, are natural monopolies. There may be many network companies, but they are all monopolies in their own geographical area. Following the neo-classical microeconomic theory, these monopolistic networks need to be regulated to prevent the network operators to seek monopoly rents, which they could do by increasing network charges above the level required to cover their costs. Regulation of charges, revenues or profits aims to achieve two goals: first, promotion of competition on the network (generation and retail) and second, protection of the consumer and economic welfare. At the same time, the regulatory framework must consider the following constraints:

- Regulated charges should be sufficient to allow full cost recovery and thereby allow adequate new investment.
- The framework should set incentives for the network operators to maintain and improve the efficiency of production.
- The framework should set incentives to create new value, where this is beneficial for society.

The latter constraint points to a new development in regulation practice and theory. Regulation already saw a major paradigm shift in the 1980s, with the shift from cost-based approaches to price-based regulation. Currently, we may face a next step: output-oriented regulation. In this section, we will explore this development.

3.1 The Regulatory Lag and Regulatory Review

In essence, regulation allows revenues to match total cost plus a fair and reasonable rate of return on capital. However, as we will explain below, if costs are simply passed

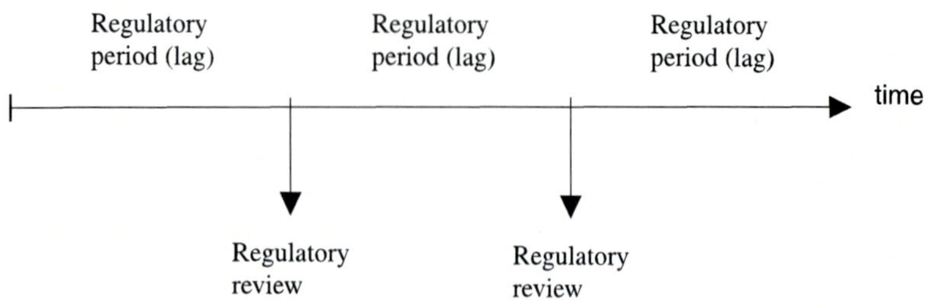

Fig. 3 Timeline of the regulatory lag and regulatory review

through to the customer then the network operator will have low incentives to keep costs low. Therefore, regulation tried to de-link allowed revenues from the firm's own underlying costs: in that case, movements in the costs do not affect allowed revenues (and prices) and therefore the firm will have an incentive to keep the costs down. The period in which allowed revenues are delinked from underlying costs is called regulatory lag or regulatory period. On the other hand, if the allowed revenues have no relation to the underlying costs, the outcome may become unreasonable; either the firm makes excessive profits or losses. Therefore, at times, the regulation needs to re-link allowed revenues to the cost base. We call this the regulatory review (in the US, this is often called rate hearing). The regulatory timeline in Fig. 3 illustrates this.

The basic models of cost-based and price-based regulation differ in precisely this point. In cost-based models, the regulatory lag tends to be short and endogenous: the link between allowed revenues and costs is strong. Price-based models try to de-link allowed revenues and costs explicitly: the regulatory period is relatively long and exogenously predetermined. Output-oriented models extend the price-based models and link revenues to some explicit output metric, irrespective of underlying costs, until a regulatory review.

As Joskow (1989 and 2014) convincingly points out, the different types of regulation may actually be quite similar. In practice, regulation is the sum of details and it is quite difficult to give an unambiguous label.

3.2 Cost-Based Regulation: Rate of Return Regulation

Cost-based regulation has a long tradition in monopoly regulation, especially in the US. The most well-known form of the cost-based regulation is rate of return regulation, where the regulatory cost base is the capital base. Rate of return regulation allows a 'fair' rate of return on capital employed (Joskow 1974; Joskow et al. 1989). The basic formula is

$$REV = OC + d \cdot (KT - CD) + s \cdot (KT - CD) + T \tag{1}$$

where

- REV is revenue,
- OC is operating costs,
- KT is historic capital value,
- CD is cumulative depreciation,
- T is taxes,
- d is depreciation rate,
- s is allowed rate of return on capital.

In addition, we define r as the cost of capital. The allowed rate of return s is to be determined by the regulator. If $s = r$, then the allowed rate of return would exactly match the investors' idea of the cost of capital. Usually, we would expect s to be somewhat larger than r as the bargaining outcome between the regulator and regulated firm. The allowed revenues are calculated from the regulatory cost base: the asset base, operating cost and taxes. Note that the operating costs (here OC, but usually denoted by OPEX) are a cost pass-through and do not have a mark-up.

In theory, the regulator sets s and all else follows from this. In particular, the firm would have to calculate its own prices and revenues following the costs in such a way that the rate of return does not exceed s. Therefore, if the costs change, the revenues (and prices) have to be adjusted to fulfil the regulatory constraint.

Rate of return regulation suffers from at least the following two drawbacks: A first problem is the low-powered incentives of cost-based regulation. Assume that the cost-based regulation is strict and thus the regulatory lag is zero. If the management of the firm now puts effort into cost reduction, it will have to reduce prices immediately to fulfil the regulatory constraint. The reverse argument also holds; additional costs can be passed on to consumers immediately. In both cases, we should expect that the incentives to control costs are low. This is the main argument for the shift towards price-based regulation, which sets strong incentives for efficiency improvement. We discuss price-based regulation later on in this chapter.

A second problem has come to be known as the Averch–Johnson (AJ) effect (Averch and Johnson 1962), also known as gold-plating or overcapitalisation (cf. Knieps 2008, for a formal exposition). The AJ-effect is typical for the rate of return regulation and does not apply to cost-based regulation in general. The rate of return regulation restricts the rate of return on capital employed, while operating expenditure is subject to a straightforward cost-pass-through. If $s > r$, it pays off to inflate the capital base at the expense of operating costs, because the capital base determines allowed profits. The inefficiency lies in the distorted ratio of CAPEX versus OPEX, also called a CAPEX-OPEX-incentive-bias (CAPEX-bias).

The AJ-effect is well established in the literature; yet, empirically it is controversial and it has not been convincingly shown to exist (see e.g., Borrmann and Finsinger 1999). Perhaps due to the partial replacement of rate of return approaches by price-based models, the AJ-effect lost popularity and did not play a major role in the regulatory debate throughout the 1990s until about 2010. Recently, the CAPEX-OPEX-incentive-bias (short, CAPEX-bias) is back in the regulatory debate of, among other things, electricity networks. First, smart grids typically rely on OPEX measures (e.g.,

IT expertise, software, curtailment, demand response, etc.). A CAPEX-bias towards traditional network assets (e.g., network expansion) would thus hamper the development of smart grids in favour of non-smart network investment. Second, network operators are increasingly facing OPEX-related tasks, e.g., congestion management which increases with renewable energies and drives the need to extend the workforce for this task.

In general, there are three sets of potential sources of CAPEX-bias: First, a CAPEX advantage, especially that the rate of return is higher than the actual cost of capital, i.e., $s > r$. Second, an OPEX disadvantage; here one can especially think of an OPEX-risk that is not fully reflected in the regulation (Brunekreeft and Rammerstorfer 2021). Third, sources for a CAPEX-bias can be caused by details in the specific regulation; this is context-dependent and differs for each country.

The debate on the CAPEX-bias received new attention with an investigation of the UK water regulator Ofwat and energy regulator Ofgem around 2011. Addressing the issue these regulators developed a variation of total expenditure (TOTEX) regulation (OFWAT 2011; OFGEM 2017; Oxera 2014). The idea is elegantly simple. A predefined fixed part of OPEX is activated and treated like CAPEX: a "fixed-OPEX-CAPEX-share (FOCS)". Under FOCS, all expenditures, whether for capital goods (CAPEX) or operational measures (OPEX), are treated equally as TOTEX. A fixed share, the capitalization rate cr, of this TOTEX is then "capitalized" (quasi-CAPEX) and the remaining part (here: $1 - cr$) is volatilized as quasi-OPEX ("pay-as-you-go"). This capitalization rate is given: fixed-OPEX-CAPEX-share. In the regulation, the resulting quasi-CAPEX and quasi-OPEX are treated in exactly the same way as the CAPEX and OPEX in the normal system. The quasi-CAPEX go into the regulatory capital base and generate depreciation and interest. The quasi-OPEX are booked within the book year. This way, ceteris paribus the firm is actually indifferent between CAPEX and OPEX and thus the CAPEX-bias is internalized.

3.3 *Price-Based Regulation: RPI-X*

In 1983, Professor Stephan Littlechild was asked by the British government to assess different regulatory regimes for the regulation of British Telecom, which was then to be liberalized and privatized. This resulted in what is now seen as a paradigm shift. Littlechild was quite critical of cost-based approaches and suggested price-based models instead (Littlechild 1983). The British government followed this advice and implemented what came to be known as RPI-X regulation (or, price-cap regulation). Soon afterwards, price-based models gained popularity in both practice and theory. The literature on price-cap regulation is vast. In the USA, price-based models are often called performance-based regulation (PBR) (NREL 2017).

As Beesley and Littlechild (1989) point out, the main reason for price-based models are high-powered incentives to reduce costs: hence, the expression incentive-

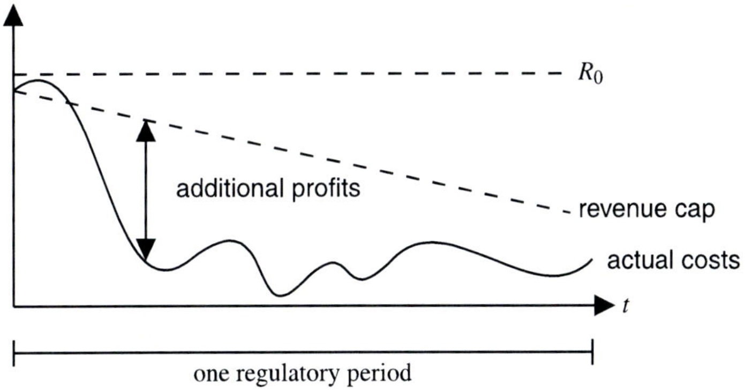

Fig. 4 Mechanics of price-based regulation

based regulation is used frequently to describe this regulation model.[2] The key point of price-based models is to explicitly de-link allowed revenues from underlying costs. This mimics a competitive outcome as prices in a competitive setting are determined by the market (i.e., demand and the supply of all firms) and not by the individual firm.

The theoretically purest form of price-based regulation is the 'tariff-basket' (Cowan 1997). In practice, we find many variations of the tariff basket. A prominent variation is the revenue cap, as described as follows:

$$\sum_{i=1}^{n} p_{i,t} \cdot Q_{i,t} \leq R_0 \cdot (1 + RPI - X) \tag{2}$$

where $p_{i,t}$ is the price and $Q_{i,t}$ is the quantity of good $i = 1, \ldots, n$ in period t, R_0 is some initial level of revenues in starting year 0, RPI is the retail price index (say, general inflation) and X is the (estimated ex-ante) expected productivity increase. The periods t are normally years within a regulatory control period of 3–5 years (see Sect. 3.1). This rule is applied within the control period and all variables change accordingly, but the rule itself is not changed. Revenues should follow this rule. Only at a regulatory review, which takes place at a predetermined moment, can the rule itself be changed.

De-linking allowed revenues from underlying own costs implies high-powered incentives to reduce costs. If the regulated firm manages to reduce its cost during the control period by more than what is expressed in X, it does not have to reduce prices for the additional cost reduction, but can instead keep these profits. This is precisely what sets the incentives to reduce costs in the first place (see Fig. 4).

At the latest at the review moment, allowed revenues are brought back to underlying own costs. In other words, allowed revenues are hardly ever completely delinked

[2] The term is somewhat unfortunate, as it is a misnomer. All regulatory mechanisms set incentives one way or another, and thus the term incentive regulation lacks meaning.

from own costs. The reason for having review periods and trying to relate revenues to underlying costs is straightforward. The result of complete de-linking may be unreasonable. Either it may turn out that the allowed prices are actually far too high, which questions the effectiveness of the regulation, or, what is worse, the allowed prices may be too low to recover full costs or warrant new investment. The review period gives the regulator the possibility of controlling the degree of reasonableness. However, if the revenues are re-aligned to underlying costs, the incentive problems mentioned above re-emerge: firms will try to game the regulation and inflate the cost base.

The high-powered incentives to reduce costs are well established. But there is a downside: what if costs go up? More precisely, price-based models work well to bring costs down, but have difficulty with cost-increasing investment. Theory is quite ambiguous about this, but in practice, we observe that regulators have started to adjust the regulatory models to facilitate more network investment (Brunekreeft and Meyer 2016). The energy transition increases the need for network investment, hence regulators have started to acknowledge this changing environment of network operators. The case of Germany illustrates this well. After a long debate about the low investment incentives of a pure revenue cap, the regulator changed the regulatory model: basically, the system was changed to an annual adjustment of capital costs, basically thus lifting the regulatory lag for CAPEX.[3] We observe similar moves in other countries. Perhaps the best example is the UK, where after 20 years of RPI-X regulation the system was replaced by a new system.

3.4 New Developments: Output-Oriented Regulation

A new development is about to emerge: output-oriented regulation, which supplements the base incentive regulation with revenue elements that reflect the achievement of specifically determined regulatory output targets or performance. Output can be anything and is broader than efficiency only. Output-oriented regulation can incentivize activities that require cost-increases and upfront expenditures and can capture external effects. We should stress that the main idea is to retain a revenue cap in the regulatory core, but supplemented with output-oriented components.

Four effects drive the development towards output-oriented regulation. First, triggered by the energy transition, network costs are increasing; the efficiency-oriented regulation is not well equipped to deal with increasing costs. Second, as pointed out by Poudineh et al. (2020), innovative activities, which have gained importance recently, face higher risks than conventional network activities. For risk-averse companies, the higher risk profile requires a move away from types of regulation (such as pure efficiency-oriented incentive regulation), which allocate a large part of the risk to the company. Instead, risky innovation activities require lower-risk types of

[3] Somewhat confusingly, it is still called "incentive regulation", although annual cost adjustment is clearly cost-based regulation.

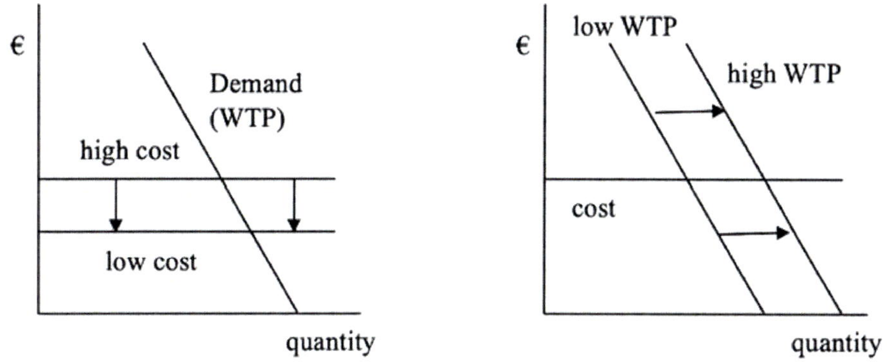

Fig. 5 A shift of the cost curve versus a shift of the demand curve

regulation. Output-oriented regulation can balance risks between pure cost-based and price-based approaches. Third, the development and the promotion of competition has resulted in a fragmented sector. Whereas competition is unquestionably beneficial, the drawback of a fragmented sector is a loss of coordination between different actors in the sector and a lack of whole system optimization. European regulators have picked this up and call this a "whole system approach" to reflect external effects (CEER 2017b). Fourth, in practice, most regulatory models do not incentivize the development of new tasks and business models (value creation). A rationale for value-creating output-oriented regulation was (unintentionally) provided in the seminal work on quality regulation by (Spence 1975, p. 420) where he notes: "Of somewhat less interest is the case where price is fixed or taken as given. In that case, the firm always sets quality too low." To see this, note the difference between a shift of the cost curve and a shift of the demand curve (Fig. 5).

In the case of improving efficiency, the cost curve goes down, while the demand stays constant. This is what price-based models are aiming at. Things change if the demand curve shifts out: by some innovation, the product is improved, such that the willingness to pay by the consumers increases. As pointed out by Spence, price-based models, where "price is fixed," cannot deal with this situation very well. As the demand curve is shifted out, additional surplus is created (the additional area under the demand curve): "value creation." If the regulation fixes the prices, the firm cannot sufficiently recoup the additional surplus and will underinvest in product improvement. This holds irrespective of whether or not the costs increase, but the problem gets worse if costs increase. This is precisely where output-oriented regulation steps in: output-oriented regulation attempts to define and quantify the product improvement (the shift of the demand curve) by some metric and link the additional consumer surplus to additional profit for the firm and thereby set the incentives for additional value creation.

Clearly, it is challenging to set the right incentives for the economically optimal outcome. First, which outputs qualify to be incentivized? Second, which metrics should we use? Third, which incentive mechanism should we use? Fourth, how

Table 1 Output categories under RIIO in the UK (OFGEM 2010)

Output categories		
Customer Satisfaction Satisfaction of consumers, including a broad spectrum of network users, with services	Reliability and availability Aspects of reliability and availability of network services that consumers are concerned with (e.g., number and duration of outages, constraints costs)	Safety Compliance with Health and Safety Executive safety standards
Condition for connection The process for new/enhanced connections to the network	Environmental impact Impact of network operations on the environment (including noise/visual impacts) and contribution to environmental targets	Social obligations Service to fuel poor and vulnerable consumers in line with Government requirements

strong should the incentives be? These and other design questions are currently in the development process. A much-quoted output-driven regulatory system is RIIO in the UK (OFGEM 2010). These main groups are safety, environmental impact, customer satisfaction, social obligations, connections, reliability and availability. Note in particular customer satisfaction, which is an intriguing output-indicator. These groups are then broken down into subgroups of more tangible outputs with specific metrics, as summarized in Table 1.

In the US, beyond broad-based PBR, regulators find it useful to strengthen incentives in pre-specified targeted goals; these are called targeted performance incentive mechanism (PIM). NREL (2017) provides the following definition for these PIMs:

"PIMs are a component of a PBR that adopts specific performance metrics, targets, or incentives to affect desired utility performance that represents the priorities of the jurisdiction. PIMs can be specific performance metrics, targets, or incentives that lead to an increment or decrement of revenues or earnings around an authorized rate of return to strengthen performance in target areas that represent the priorities of the jurisdiction."

NREL (2017, p. xii and pp.61) provides a long list of PBRs and PIMs being used operation in the US electric utility industry. To mention a few which are or can be of interest for the network operator:

- Incentives for implementation of renewable energies,
- Renewable energy performance metrics,
- Operational incentives: improved interconnection request response times,
- Operational metrics: incentives to improve reliability,
- Incentives to support competition.

Pfeifenberger (2010) mentions inter alia the following PIMs relevant for the network:

- "External" system costs (losses, congestion, ancillary services)

- Infrastructure investments (mains replacement, transmission, renewables)
- Non-cost goals: reliability, service quality, end-use efficiency (DSM)

At the moment, the development of output-oriented regulation is just starting and many implementation issues still have to be addressed and are issues for further research.

4 Grid and Market Interface: Network Congestion Management Through Pricing of Grid Use and Decentralized Approaches

In the majority of electricity markets, the price only reflects the costs of producing electricity and excludes the cost of transportation and distribution. Figuratively speaking, electricity is traded as if it could be delivered anywhere at any time. This approach is unproblematic as long as electricity grids are not congested, which has been mostly the case for a long time. When many users withdraw or feed-in electricity simultaneously, however, the technical limits of the network can be reached, i.e., the network becomes congested and cannot be expanded easily. In the distribution grids, this issue is becoming increasingly important due to the increase of decentralized renewable generation (e.g., simultaneous feed-in from wind power plants) and new electric devices (e.g., simultaneous charging of electric vehicles). To address this challenge, network congestion management is needed to set signals for optimal short-term network use and optimal long-term investment. In the following, two general approaches of network congestion management are outlined that address the interface of the electricity grid and market[4] and that are currently undergoing changes due to the development of smart grids: first, network pricing and, second, decentralized congestion management.

4.1 Network Pricing: Principles, Designs and Smart Grid Developments

Being unbundled from generation and classifying as a natural monopoly, the transmission and distribution of electricity are typically priced separately and prices are regulated to prevent the monopolists from abusing their power. One purpose of charging network usage is to finance the networks. In the previous section, different regulatory approaches have been explained that can be used to determine the level of allowed revenue of the networks to achieve this aim. Another reason why network users pay for network usage is to set signals for efficient short-term network use and

[4] Another approach is integrated energy and network pricing known as nodal pricing or locational marginal pricing. For more details on locational marginal pricing, see e.g., Stoft (2002).

long-term development of the network. This purpose has become more relevant in recent years due to increasing network congestions. At the same time, the development of smart grids, giving real-time information on network usage, promises a new approach to this challenge.

Further principles for tariff design, as laid out by Bonbright et al. (1988), that are considered by regulators also include, e.g., fairness, predictability and stability. More recently, the Council of European Regulators (CEER) identified the following seven principles for the design of distribution network tariffs (CEER 2017a):

1. Cost reflectivity,
2. Non-distortion,
3. Cost recovery,
4. Non-discrimination,
5. Transparency,
6. Predictability,
7. Simplicity.

Ensuring cost reflectivity and cost recovery alone is already non-trivial when it comes to electricity networks. In the following, the first theoretical approaches and current practices of network pricing are outlined. Then new developments triggered by network congestions and smart grids are described.

Welfare maximization implies that the price of a good, here network use, should equal its marginal costs (first-best allocation) (Hotelling 1938). While ensuring allocative efficiency, marginal cost pricing of electricity networks has one major drawback: due to large fixed costs of electricity networks, marginal costs are below average costs. Marginal costs pricing would, therefore, not enable the network operator to cover her costs as illustrated in Fig. 6.

The remaining part of the costs needs to be covered differently, e.g., by an additional charge. This approach finds expression in a two-part tariff: The first part of the tariff is a charge per unit of energy (kWh) accessed from or feed-in to the grid that

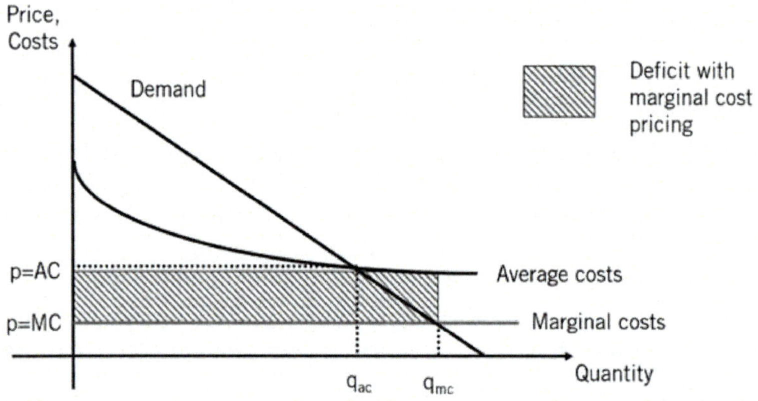

Fig. 6 Drawback of marginal cost pricing when average costs exceed marginal costs

equals marginal cost. The second part is a charge that is independent of consumption or production, which serves to cover the remaining (fixed) costs.

Based on the approach of marginal cost pricing, but considering the constraint of breaking even is the so-called Ramsey–Boiteux pricing[5] (or short: Ramsey pricing). Mathematically speaking, prices are determined by maximizing welfare subject to the constraint that total costs are covered. Central to this concept is that users are being discriminated according to their price elasticity of demand, i.e., how strongly consumers react to a change in price. The resulting price equals marginal costs plus a surcharge, which depends on the inverse of the price elasticity of demand. This means that the surcharge increases with a decrease in the elasticity of demand (see, e.g., Borrmann and Finsinger 1999, for more details on Ramsey pricing). In other words, the more the users depend on electricity and cannot switch to other commodities, the more they pay. Those who can "run away" pay less.

Related to this approach, and partly developed by the same scholars, is the concept of peak-load pricing. The starting point for this concept is that demand for electricity and the use of the network varies over different periods (e.g., seasons or during a day or week), while the network capacity remains almost constant. Peak demand then determines the required network capacity and hence the total costs that need to be recovered. During off-peak periods the installed capacity is often under-utilized. To signal network scarcity and to prevent excess capacity, network charges need to be adjusted accordingly for each period. This is addressed in the peak-load pricing approach as follows[6]: Consider a day divided into two periods: a peak-load period and an off-peak period. Once installed, generation capacity can be used in both periods. Constant operating costs are incurred per unit per period and constant marginal capacity costs per unit of capacity. The problem is to determine the optimal output per period and the corresponding prices. The maximum output in either period determines the required capacity. If off-peak demand is very low and would at no price fully utilize the network capacity, it should only be charged operating costs while peak demand covers the entire capacity costs. Yet, if, also, during the off-peak period network capacity is fully utilized, then off-peak demand should also contribute to covering capacity costs, but only according to its willingness to pay for this capacity (see Crew et al. 1995, for more details on peak-load pricing).

In practice, tariffs often vary for different network user groups. First, electricity consumers and generators are typically treated differently, i.e., the latter are often exempted from network charges (e.g., in Germany or the Netherlands) or incur only a smaller share of network costs (e.g., 3% in France, 38% in Sweden).[7]

Second, consumer tariffs often differ for the groups of large industrial consumers (connected to the transmission grid), commercial and small industrial consumers

[5] The approach is named after Frank Plumpton Ramsey and Marcel Boiteux: Ramsey (1927) published the result first in the context of optimal rates of taxation; Boiteux (1956) applied it to public monopolies. Baumol and Bradford (1970) synthesized the approach.

[6] Steiner (1957) addressed the peak-load problem graphically; the example given here is taken from his publication.

[7] ENTSO-E Overview of Transmission Tariffs in Europe: Synthesis 2018. Online available at: docstore.entsoe.eu/Documents/MC%20documents/TTO_Synthesis_2018.pdf.

(connected to the distribution grid), and residential consumers. For historic and other reasons (e.g., traceability, simplicity, etc.) the two- (or three-) part tariff, developed in the late nineteenth century, is still the dominant pricing structure for residential consumers in many countries, e.g., the US, Australia, Great Britain, Germany, etc. One part of the tariff is fixed, while the other is energy-based (per kWh use) and/or capacity-based (e.g., per kW connected). Depending on the proportion covered by the fixed part, the incentives for efficient grid use are rather low. In contrast, a fixed component for non-domestic consumers connected to the distribution grid is rare and even more for large consumers connected to the transmission grid (exceptional cases in the EU for the latter are, e.g., France and Sweden) (AF-Mercados 2015). The increase of decentralized electricity generation and self-supply has changed the network utilization in particular on the low voltage level. With a future increase of flexibility providers (e.g., decentralized storage) and new flexible electric devices (e.g., electric vehicles and heat pumps) new challenges will arise for the distribution networks. This development also changes the focus of network tariff design. While historically, the focus lay on the efficient allocation of fixed costs of (oversized) network capacities, in the light of network congestion and the need for network expansion the focus shifts towards incentives for network cost reducing deployment and operation. The current developments trigger a redesign of network tariff structures for users connected to the distribution grids. First, network charges may become more based on capacity than on net consumption also for small network users to prevent disincentives for households with self-supply. Second, the implementation of smart meters enables dynamic time differentiation of tariffs users and even real-time pricing for households and other small network users, which, in turn, signals actual network scarcity.

Time-varying network pricing can come in different degrees and varies from time of use (ToU) to real-time pricing (RTP). ToU pricing implies that the network charges vary for specific predefined time periods and are highest during times of expected peak demand. Several countries have implemented ToU tariffs to deal with the peak-load problem. In Europe, ToU tariffs vary between one differentiation, e.g., day and night (e.g., in Finland, Greece), up to four-time differentiated tariffs (in Northern Ireland), whereas no ToU signals are used in, e.g., Germany, Italy and Sweden (CEPA 2015).

Real-time pricing in contrast reflects the actual network usage and scarcity in real-time or at least with a much higher temporal granularity (e.g., 15 min) than ToU tariffs. Real-time network pricing has, hence, the undisputed potential to significantly increase the efficiency in network use. Concerning other principles of network pricing as outlined above, it has, however, also several shortcomings. Its dynamic nature, for instance, decreases the predictability and stability of tariffs for the customers. The uncertainty also increases for the utilities in two regards: first, the effect on the network use is not known and, second, the revenue becomes less predictable. This explains why, e.g., the German regulatory agency opposes dynamic network pricing on the basis of transparency and uncertainty (Bundesnetzagentur 2015). A way forward in this regard may be to limit the dynamics of the network pricing to certain corridors and the lead times of price changes, as well as introducing optional instead of obligatory participation in dynamic pricing (on the latter, see, e.g., Borenstein 2013).

4.2 Decentralized Congestion Management

Since mostly, energy pricing does not integrate network pricing, network congestion is managed by the network operator. In the long term, he/she can resolve network congestions by network expansion or upgrades. In the short term, she can undertake operational measures in order to safeguard the operational limits of the electricity grid. One of these measures is the so-called redispatch: After the closure of the energy market, the network operator examines whether the single feed-ins and withdrawals determined by the market (i.e., the dispatch) are technically feasible. If this is not the case, she can change the geographical distribution of generation and/or load patterns in order to reduce the load flow on the congested lines. To illustrate this, consider a simple network with two nodes A & B. The transmission line connecting the two nodes has a capacity of 50 MW. A generator is connected to each node with a capacity of 100 MW each. Generator A has marginal production costs (MC_A) of 50 EUR/MWh, generator B (MC_B) of 80 EUR/MWh. Load is only connected to node B and demands 100 MW (L_B) for a certain hour. Energy is traded in a single market disregarding the transmission limits. In this setting, generator A would completely cover demand for that hour ($G_A = 100$ MW; $G_B = 0$ MW). The result of the energy market is shown in the left part of Fig. 7. Of the 100 MW sold, however, only 50 MW can be transmitted to node B. To stay within the technical limits, generator A needs to reduce her production by 50 MW (downward redispatch). At the same time, generator B needs to increase production by 50 MW (upward redispatch) to keep the energy balance.[8] The result of the redispatch is illustrated in the right part of Fig. 7.

Historically, power plants were predominantly large in scale and connected to the high-voltage level. That is why, still today, congestion management falls into the responsibility of the transmission system operators and is mostly conducted on the high-voltage level. The past years, however, have seen an increase in decentralized (renewable) power production connected to the lower-voltage levels. Furthermore, load connected to the distribution grid is becoming more flexible, e.g., electric vehicles or battery storage units. This has changed the network utilization, in particular, on the low voltage level and the still ongoing development is likely to put more pressure on distribution grids in the future. Against this backdrop, decentralized congestion management has gained attention in recent years as a means to integrate renewable generation and flexible demand into congestion management, as well as to address regional network congestions. The development of smart grids supports this approach.

[8] Alternatively, load could reduce withdrawal by 50 MW.

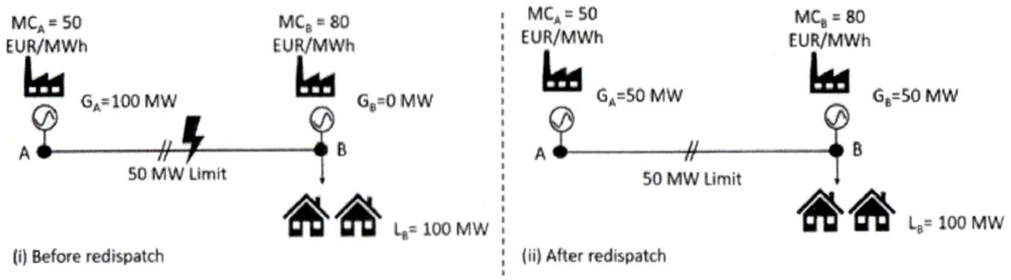

Fig. 7 Illustration of (i) a situation with a technically infeasible energy market result before redispatch and (ii) after it is solved in redispatch in a two-node example

Currently, the design and implementation of decentralized congestion management schemes is being debated. Many options are in the discussion. Essential design elements are

- the way network congestion is communicated and how flexibility is demanded,
- the determination of compensation,
- the implementation.

Figure 8 outlines potential forms of decentralized congestion management based on these characteristics.

The network operator can request flexibility (i.e., deviation from the energy market result) passively by means of a quota. Depending on the congestion situation, i.e., over- or undersupply, all generators or all loads within the congestion region are apportioned until the network limits are reached. Alternatively, the network operator can actively demand flexibility from the local loads or generators.

The compensation for participating in redispatch can be market-based or regulated (cost-based). In the first case, supply and demand of flexibility determine the price. In the case of regulated redispatch, the compensation is determined by the regulator and can be cost-based, derived from markets or a reduction in network charges. The idea behind regulated or cost-based redispatch is to keep the electricity market and network separate from one another and to prevent the abuse of potential market power of congestion critical power plants. Redispatched units are to be placed economically in the same position as they would have been without the network operators interference, making them indifferent to redispatch. Thus feedbacks to the electricity markets shall be prevented. In Switzerland and Germany, for example, compensation for centralized redispatch is cost based, i.e., the generators are compensated for the incurred costs of upward redispatch and foregone profits in the case of downward redispatch. In Spain, transmission constraints are solved based on the bids previously committed to the energy market.

Market-based redispatch can be implemented in a separate dedicated market platform, where the network operator is the single buyer. Alternatively, an existing energy market, e.g., the intraday market, can be adjusted with locational information so as to also serve for redispatch. In the UK, for instance, the balancing market is used for

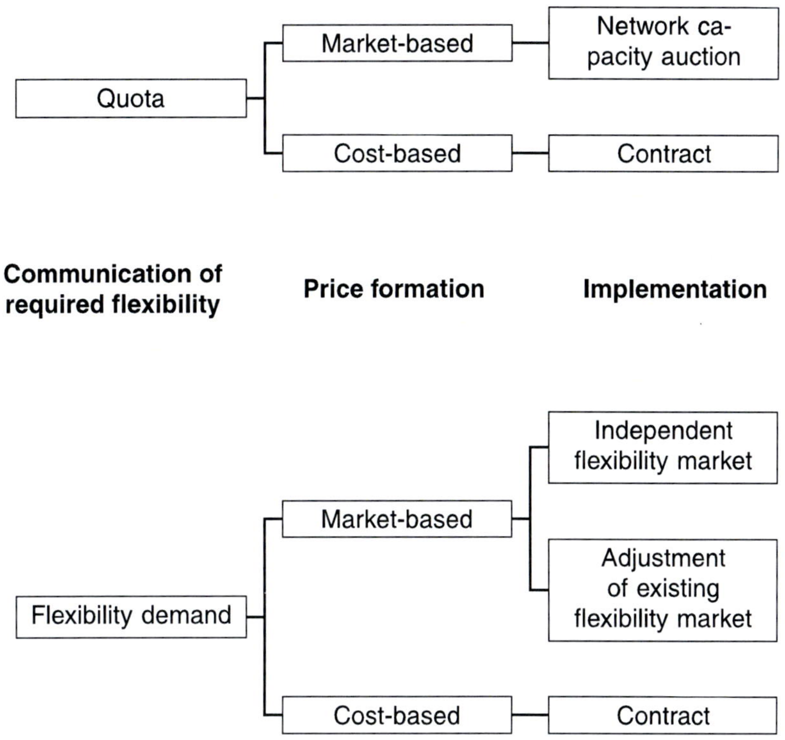

Fig. 8 Overview on potential forms of decentralized congestion management. (based on Ecofys and IWES 2017)

centralized congestion management. Regulated redispatch is handled in a contract between network operator and flexibility provider.

New decentralized redispatch market platforms are currently tested in pilot projects in different European countries (for an overview, see, e.g., Schittekatte and Meeus 2019; Radecke et al. 2019). Whether decentralized congestion management can effectively and efficiently integrate flexible network users and relieve regional network congestions remains to be seen.

5 Summary

This chapter explored regulatory issues of smart grids. We have tried to strike a balance between the basics of such regulation principles and topical developments, which are triggered by the development towards smart grids. The chapter focussed on three fields. First, an electricity sector is a complex value chain with monopoly networks as platforms between commercial market areas. Various governance structures (commonly known as network unbundling) attempt to secure non-discriminatory access to the monopoly networks for all eligible parties in order to promote a competitive level-playing field. This debate is now emerging in a new dress for the distribution networks following the development of smart grids. Second, we describe network regulation, meaning the regulation of revenues and profits of the monopoly networks. We discuss different approaches (cost-based versus price-based) and, more topical, a new development towards output-oriented regulation. The latter is triggered by emerging new tasks for the network operators, which create value but are not very well facilitated under standard regulatory models. Third, we discuss the interface between the network and the market. More precisely, we discuss network charging to address network congestion. The latest development is how to adequately address decentralized network congestion: the discussion is whether we can design functioning flexibility markets.

Governance With smart grids, the distribution grid operators' role extends into the competitive realm of the electricity supply chain. The increasing interaction between the monopolistic networks and market activities in smart grids drives a discussion about the need for further unbundling measures. Currently, legal unbundling is the norm in Europe, but deemed to be potentially insufficient to facilitate smart grids. Ownership unbundling and the introduction of IDSO come with too high costs that might even exceed the benefits of further unbundling. Hence, the application of the IDO, an adaptation of the ITO successfully applied on the transmission grid level in Europe, might provide a balanced approach to secure a level-playing field (i.e., competition) in smart grids and at the same time a sufficient level of coordination between the networks and its users.

Network regulation Regulation of the revenues or profits of the monopoly network aims to protect consumers and competition. The two big variations are cost-based models (in particular, rate of return regulation) and price-based models (e.g., RPI-X). The key drawback of cost-based models is the low incentive to reduce costs. In reverse, this is the main advantage of price-based models: it sets strong incentives to reduce costs and increase efficiency. However, price-based models are not well equipped to facilitate cost-increasing activities which create new value. This is where output-oriented regulation comes in: output-oriented components supplement the base regulatory model with revenue elements that reflect the achievement of specifically determined regulatory output targets or performance.

Interface between network and market Electricity is predominantly traded as though it could be transported without constraints. When too much electricity is fed-in or withdrawn simultaneously, however, the network can become congested. Ensuring the efficient use of the grid is the aim of network pricing. Different network pricing approaches have been developed to signal network scarcity. In practice, however, network pricing schemes give little or no incentives for efficient network use in particular in the distribution grids. Smart grids, giving real-time information on network use, can improve network pricing at the distribution level in this regard. Apart from network pricing, decentralized congestion management based on smart grids can be used to deal with network congestion at the distribution level. Several designs for decentralized congestion management are currently tested.

Review Question

- Unbundling results in a trade-off between coordination and competition. With smart grids, the balance between these two on the distribution grid level changes. Why?,
- Which unbundling regime on the distribution grid level is most suitable to accelerate the diffusion of smart grid technologies? Please elaborate.,
- An important assumption for the so-called Averch–Johnson effect is that the allowed rate of return on capital is higher than the true cost of capital: $s > r$. Explain the "Averch–Johnson effect." Discuss why this is assumption important and the plausibility of the assumption.,
- Explain what output-oriented regulation is and what it tries to achieve. Briefly discuss an example of output-oriented regulation in the context of smart electricity grids.,
- The extensive roll-out of smart devices in the electricity grid (e.g., smart meters) may enable a more efficient use of the grid. Explain how this may be the case in the context of network pricing.,
- With regard to prevalent network pricing principles, real-time pricing of network usage has several shortcomings for network users and network operators. Discuss.

References

AF-Mercados E (2015) Study on tariff design for distribution systems. Technical report, Technical Report. https://ec.europa.eu/energy/sites/ener/files

Averch H, Johnson LL (1962) Behavior of the firm under regulatory constraint. Am Econ Rev 52(5):1052–1069

Balmert D, Brunekreeft G (2010) Deep-isos and network investment. Compet Regul Netw Ind 11(1):27–49

Baumol WJ (1977) On the proper cost tests for natural monopoly in a multiproduct industry. Am Econ Rev 67(5):809–822

Baumol WJ, Bradford DF (1970) Optimal departures from marginal cost pricing. Am Econ Rev 60(3):265–283

Beesley ME, Littlechild SC (1989) The regulation of privatized monopolies in the united kingdom. RAND J Econ, pp 454–472

Boiteux M (1956) Sur la gestion des monopoles publics astreints à l'équilibre budgétaire. Econom J Econom Soc, pp 22–40

Bonbright JC, Danielsen AL, Kamerschen DR (1988) Principles of public utility rates. Columbia University Press, New York

Borenstein S (2013) Effective and equitable adoption of opt-in residential dynamic electricity pricing. Rev Ind Organ 42(2):127–160

Borrmann J, Finsinger J (1999) Markt und Regulierung. Vahlen München

Brunekreeft G (2015) Network unbundling and flawed coordination: experience from the electricity sector. Util Policy 34:11–18

Brunekreeft G, Goto M, Meyer R, Maruyama M, Hattori T (2014) Unbundling of electricity transmission system operators in germany: an experience report. Technical report, Bremen Energy Working Papers
Brunekreeft G, Keller K (2001) Sektorspezifische ex-ante-regulierung der deutschen stromwirtschaft? Technical report, Diskussionsbeiträge//Institut für Verkehrswissenschaft und Regionalpolitik
Brunekreeft G, Meyer R (2016) Anreizregulierung bei Stromverteilnetzen: Effizienz versus Investitionen. Perspektiven der Wirtschaftspolitik 17(2):172–187
Brunekreeft G, Rammerstorfer M (2021) Opex-risk as a source of capex-bias in monopoly regulation. Compet Regul Netw Ind 22(1):20–34
Bundesnetzagentur (2015) Bericht der Bundesnetzagentur zur Netzentgeltsystematik Elektrizitüt
Burger S, Jenkins JD, Batlle López C, Pérez Arriaga JI (2018) Restructuring revisited: competition and coordination in electricity distribution systems
CEC (2014) Commission staff working document, report on the ito model
CEER (2016) Review of current and future data management models. Technical report. Ref: C16-RMF-89-03
CEER (2017a) Electricity distribution network tariffs ceer guidelines of good practice. Technical report. Ref: C16-DS-27-03
CEER (2017b) Incentives schemes for regulating dsos, including for innovation. Technical report. Ref: C16-DS-28-03
CEPA (2015) European transmission tariff structure. presentation acer workshop on electricity transmission tariff harmonization
Coase RH (1937) The nature of the firm. Economica 4(16):386–405
Cowan S (1997) Price-cap regulation and inefficiency in relative pricing. J Regul Econ 12(1):53–70
Crew MA, Fernando CS, Kleindorfer PR (1995) The theory of peak-load pricing: a survey. J Regul Econ 8(3):215–248
De Nooij M, Baarsma B (2009) Divorce comes at a price: an ex ante welfare analysis of ownership unbundling of the distribution and commercial companies in the dutch energy sector. Energy Policy 37(12):5449–5458
Ecofys and IWES (2017) Smart-Market-Design in deutschen Verteilnetzen: Entwicklung und Bewertung von Smart Markets und Ableitung einer Regulatory Roadmap. Studie im Auftrag von Agora Energiewende
European Commission (1996) Directive 96/92/CE of the European Parliament and of the Council of December 19, 1996. Concerning Common Rules for the Internal Market in Electricity, Technical report
European Commission (2003) Directive 2003/54/EC of the European Parliament and of the Council of June 26, 2003, Concerning Common Rules for the Internal Market in Electricity and Repealing Directive 96/92/EC. Technical report
European Commission (2009) Directive 2009/72/EC of the European Parliament and of the Council of 13 July 2009 Concerning Common Rules for the Internal Market in Electricity and Repealing Directive 2003/54/EC. Technical report
European Commission (2011) Energy markets in the european union 2011. Technical report
European Commission (2019) Directive 2019/944 of the European Parliament and of the Council of 5 June 2019 on common rules for the internal market for electricity and amending Directive 2012/27/EU. Technical report
Friedrichsen N (2015) Governing smart grids: the case for an independent system operator. Eur J Law Econ 39(3):553–572
Hotelling H (1938) The general welfare in relation to problems of taxation and of railway and utility rates. Econom J Econom Soc, pp 242–269
Joskow PL (1974) Inflation and environmental concern: structural change in the process of public utility price regulation. J Law Econ 17(2):291–327
Joskow PL (1996) Introducing competition into regulated network industries: from hierarchies to markets in electricity. Ind Corp Chang 5(2):341–382

Joskow PL (2014) Incentive regulation in theory and practice: electricity distribution and transmission networks. In: Economic regulation and its reform: what have we learned?, pp 291–344. University of Chicago Press
Joskow PL, Bohi DR, Gollop FM (1989) Regulatory failure, regulatory reform, and structural change in the electrical power industry. Brook Pap Econ Act. Microecon 1989:125–208
Joskow PL, Schmalensee R (1983) Markets for power: an analysis of electric power deregulation. MIT Press, Cambridge, MA
Joskow PL, Schmalensee R (1983b) Markets for power: an analysis of electrical utility deregulation. The MIT press
Knieps G (2008) Wettbewerbsökonomie: Regulierungstheorie, Industrieökonomie. Springer, Wettbewerbspolitik
Littlechild SC (1983) Regulation of British Telecommunications' profitability: report to the Secretary of State, February 1983. Department of Industry
Meyer R (2012) Vertical economies and the costs of separating electricity supply–a review of theoretical and empirical literature. Energy J 33(4)
Nillesen P, Pollitt M, et al (2019) Ownership unbundling of electricity distribution networks. Econ Energy Environ Policy 10(1)
NREL (2017) Next-generation performance-based regulation: emphasizing utility performance to unleash power sector innovation. Technical report, National Renewable Energy Lab, Golden, CO (United States)
OFGEM (2010) A new way to regulate energy networks. Final Decision, Office of Gas and Electricity Markets, London
OFGEM (2017) Riio-2 framework consultation
OFWAT (2011) Capex bias in the water and sewerage sectors in england and wales – substance, perception or myth?
Oxera (2014) Ofwat's final methodology: now for implementation
Pfeifenberger J (2010) Incentive regulation: introduction and context. Technical report, Presented at AUC PBR Workshop, Edmonton, Alberta
Pollitt MG (2012) Lessons from the history of independent system operators in the energy sector. Energy Policy 47:32–48
Poudineh R, Peng D, Mirnezami SR (2020) Innovation in regulated electricity networks: incentivising tasks with highly uncertain outcomes. Compet Regul Netw Ind 21(2):166–192
Radecke J, Hefele J, Hirth L (2019) Markets for local flexibility in distribution networks
Ramsey FP (1927) A contribution to the theory of taxation. Econ J 37(145):47–61
Schittekatte T, Meeus L (2019) Flexibility markets: Q&a with project pioneers. Robert Schuman Centre for Advanced Studies Research Paper No, RSCAS, p 39
Spence AM (1975) Monopoly, quality, and regulation. Bell J Econ, 417–429
Steiner PO (1957) Peak loads and efficient pricing. Q J Econ 71(4):585–610
Stoft S (2002) Power system economics: designing markets for electricity. Wiley-IEEE Press
Williamson OE (1979) Transaction-cost economics: the governance of contractual relations. J Law Econ 22(2):233–261

Modeling Smart Grid Systems

Dominik Möst, Hannes Hobbie, Steffi Misconel, David Schönheit, and Christoph Zöphel

Learning Objectives

- Be able to describe elements of smart grids that can be represented in models.
- Be able to classify and identify dimensions in energy system models.
- Be able to mathematically formulate smart energy system models.
- Be able to explain how the main drivers of smart grids impact model-based representations.

1 Introduction

This chapter is motivated by the transformation of the energy system toward a smart grid economy which also necessitates new solutions in the field of decision support tools that are used by system operators and market stakeholders. Trends that can be observed in the management of smart grids are an increasing orientation toward

D. Möst (✉) · H. Hobbie · S. Misconel · D. Schönheit · C. Zöphel
Chair of Energy Economics, Technische Universität Dresden, Münchner Platz 3, 01069 Dresden, Germany
e-mail: dominik.moest@tu-dresden.de

H. Hobbie
e-mail: hannes.hobbie@tu-dresden.de

S. Misconel
e-mail: steffi.misconel@tu-dresden.de

D. Schönheit
e-mail: david.schoenheit@tu-dresden.de

C. Zöphel
e-mail: christoph.zoephel@tu-dresden.de

C. Dinther et al. (eds.), *Smart Grid Economics and Management*,
Lecture Notes in Energy 51, https://doi.org/10.1007/978-3-030-84286-4_6

digital and intelligent solutions and a stronger coupling between different energy sectors as well as a growing interaction between different stakeholders. Examples include the electrification of district heating via heat pumps, mobility applications, e.g., electric vehicles, or consumers who provide energy from rooftop photovoltaic systems to grid operators to ensure grid stability enabled by smart grid devices. These developments also have implications for the model-based representation of smart grid systems.

Classic energy system models have a long history and are widely used since the two oil crises in the 1970s. Energy system models support decision-makers in questions regarding energy regulation and policies for infrastructure planning of energy generation, conversion, and transportation. Literature in the field of energy system modeling is vast and the reader of this book might ask how the modeling of smart grids differs from traditional modeling techniques. This chapter perceives a gap in the subject on model requirements that result in particular from smart grid economics and management. It provides an overview of different design aspects, challenges, and current trends associated with the model-based representation of smart grid systems, and additionally provides a detailed literature review of various modeling approaches in this research field.

The chapter is structured as follows: Sect. 2 gives an introduction to general modeling aspects of smart energy systems. For this purpose, a systematic taxonomy of smart grid systems is developed and different concepts are classified according to application scopes. Modeling of smart grid systems can be done at different scales and from various perspectives, thus different modeling approaches are introduced based on brief mathematical descriptions are presented in Sect. 3. These comprise small-, medium- and large-scale applications, bottom-up demand side models as well as bi-level approaches. Based on the structure of Sect. 3, Sect. 4 provides a comprehensive overview about literature with regard to smart grid modeling. Current trends in the modeling of smart grid systems are discussed in Sect. 5. In the final Sect. 6, a conclusion, an outlook and some exercises are presented.

2 Taxonomy and Classification of Smart Grid Systems

Modeling of energy systems has a long tradition and got a strong push with the two oil crises in the 1970s. In general, the purpose of energy system modeling and analysis is to improve and support the decision-making process in the energy sector with regard to technology choices, policies, and infrastructures for energy supply and energy conversion. Therefore, models try to consider "reality" in a systematic and knowledge-based manner. Depending on the question, a wide variety of factors and framework conditions must be considered. Developments in energy and key technologies, the limited nature of fossil resources and climate change, demographic change, the political, social, and economic framework conditions, the pursuit of sustainability—all these factors are only examples to be taken into account when analyzing and modeling an energy system. In general, energy system models can be classified in three main dimensions (see also Möst et al. 2009):

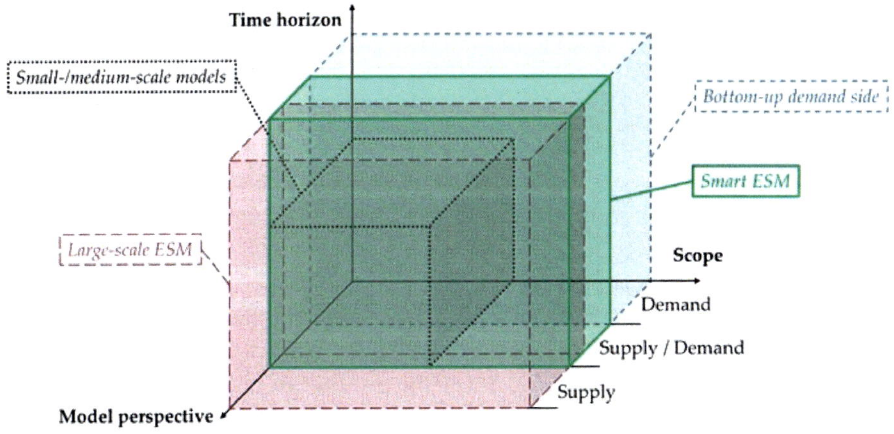

Fig. 1 Categorization of smart grid modeling approaches

- What is the energy system under consideration? The examined energy system can be the global one, the European, a national, that of a district, an industrial location, or a house. As a global energy system is not analyzed on the same level of detail as, e.g., a system on household level, system boundaries, and level of detail have to be defined.
- What time horizon and correspondingly what time resolution is addressed? The time horizon can range from a short time horizon and a high time resolution (e.g., analyzing the frequency behavior in a grid) to several decades and a lower time resolution (e.g., analyzing the development of the energy system until 2050).
- And finally, what is the model perspective? The perspective can be supply-oriented, grid-oriented, demand-oriented or even an integrated model approach.

Figure 1 depicts these three dimensions of energy system models: (1) The time horizon, describing the short-, medium- or long-term analysis, (2) the scope, which encompasses the level of detail and geographical range, (3) the model perspective, referring to a supply-oriented, demand-oriented, or integrated model approaches. This classification is also used in the course of this chapter.

It can also be applied to classic energy system models. Accordingly, the question arises what the difference to the modeling of *smart* grid systems is. "Smart" can refer to the system under consideration. Additionally, the other chapters of this book provide a good overview on what a smart energy system can entail. Modeling smart grids and the involved consumers or prosumers can be done on vastly different scales and from different perspectives. In the following, the differences in the aforementioned scope of smart grid systems are discussed by delimiting modeling approaches, although there is a fluent transition between the system and modeling boundaries.

In general, it can be stated that the main drivers of smart grid systems are the three Ds—Decarbonization, Digitalization, Decentralization—which are often considered to be the pillars of creating the green energy economy of the future. While these drivers were already introduced in chapter "Energy Systems Today and Tomorrow". However, this section explains how the three "Ds" stimulate and affect modeling.

Decarbonization
Decarbonization means that carbon-based fuels, such as oil, gas, coal, and lignite should no longer to be used for electricity generation. Renewable energy sources in particular are expected to replace these fuels and contribute to a more sustainable energy system. However, technologies with high potentials—such as wind energy and photovoltaic—are dependent on weather conditions. This creates challenges concerning the balance of electricity supply and demand. In addition, the potentials of these renewable energy sources are not necessarily close to demand, leading to longer transportation distances and hence the need for infrastructure adaptation.

Accordingly, the temporal and spatial resolution in energy system models must take into account the challenges posed by the higher share of renewable energy:

- Time component: Demand and supply have to be balanced and thus challenges in connection with the fluctuation of electricity generation (RES) and its balancing have to be addressed by smart grid modeling. In general, an hourly resolution or even a quarter-hourly resolution is state-of-the-art when modeling smart grid systems.
- Spatial component: As renewable sites are not necessarily located in close proximity to demand, transport and distribution of electricity pose new challenges for today's infrastructure. But also distribution grids are affected by new suppliers (e.g., photovoltaic) and demand technologies (e.g., electric mobility). The spatial component in high-resolution models is usually accounted for using at a NUTS3 level.[1]

In combination with new technologies on the supply and demand side, such as photovoltaic, electric mobility, heat pumps, etc., a further trend is decentralization.

Decentralization
Decentralization comes hand in hand with new technologies, providing decentralized feed-in of electricity and are intended to replace a "few" large generation plants in the long term. In consequence, electricity generation is dispersed across many smaller plants.[2] Furthermore, decentralization also refers to the increasing amount of embedded generation, for example, combined heat and power plants on industrial sites or solar panels on residential properties. As (larger) conventional power plants reach the end of their lifetimes, they are being replaced by wind farms, solar fields,

[1] Nomenclature des unités territoriales statistiques (NUTS) is a geo-code standard for referencing the subdivisions of countries for statistical purposes. NUTS3 refers to the district or municipalities level.

[2] However, this does not necessarily mean that energy based on renewable sources is always decentralized. While the power of single offshore wind energy plants is still small in comparison to large-scale fossil-fuel or nuclear-fuel based power plants, these single offshore plants are often grouped together to an entire farm resulting in connection points with large power injections (similar to the magnitude of large-scale plants).

hydropower, marine generation, and biomass, and thus decentralization is becoming more prevalent. Furthermore, sector coupling is also contributing to decentralization by providing additional electricity demand with high flexibility in general. This can help to balance supply and demand as decentralized as possible. Sector coupling is driven by so-called power-to-X (PtX) technologies, which means that electricity is used to provide energy services substituting common fossil energy carriers. Among others this is especially power-to-heat (e.g., district heating provided by heat pumps), power-to-vehicle (e.g., electric mobility) as well as power-to-chemicals, in particular power-to-gas (e.g., green hydrogen). This has given rise to so-called cellular concepts that directly address the topic of decentralization.

Decentralization poses a number of challenges, which also have to be analyzed with the modeling of smart grid systems:

- With decentralization, regional autarky and self-sufficiency are gaining in importance. Several questions in this context with regard to the level of decentralization are addressed by modeling smart grid systems.
- As renewable site potentials (e.g., wind offshore) are often far distanced from demand centers, there is still the duality of a centralized and decentralized supply of energy, resulting in several challenges for the infrastructure.
- Furthermore, interaction and participation of consumers are gaining in importance. Especially the possibility of generating electricity decentrally and controlling one's own consumption (smart demand) has led to so-called prosumers as new market participants. While, techno-economic modeling of energy systems was sufficient a few years ago, today behavioral and societal aspects pose additional challenges and barriers when it comes to making such a socio-technical transformation a reality. This requires further analysis techniques that also take behavioral and societal aspects into account. An example for considering behavior of agents is agent-based modeling, which has grown in importance.

Resulting from these developments, smart grid modeling is more specific and addresses topics at the level of households or industry site level and additionally considers societal concepts.

Finally, the last trend is digitalization.

Digitalization

With digitalization, several new possibilities and applications arise. Real-time information and control is just one example which provides new opportunities. Consequently, effective management and monitoring is essential and achievable with state-of-the-art digital technology when implemented across all areas of the electricity system, from generation to transmission, distribution, supply, and demand. The core infrastructure of the grid is still using similar switches to those used in the 1950s, and therefore requires some further upgrades in order to realize the full potential of digitalization. However, this process will steadily take place and provide new opportunities in the next years and decades. With digitalization, new technologies enter the market, which affect the energy system and thus have to be considered in modeling. Two selected examples are

- Smart metering: A smart meter is an electronic device that records consumption of electric energy and communicates the information to the electricity supplier for monitoring and billing. New concepts for controlling demand and appropriate incentive mechanisms are under development and the development is supported by analysis based on smart grid modeling.
- Smart grids[3]: The emergence of smart grids stimulated electric utilities, scientists, and vendors to develop comprehensive and sustainable solutions for the different elements of smart grids.

In general, digitalization allows for new business models behind the infrastructure, especially behind the metering devices. While traditional business models in the energy sector depend on infrastructure (large generators and power grids), which can be described by the principle of "produce big and sell small," many smart business models in the energy sector act behind the meter (e.g., smart communities, pooling of demand flexibility, electric vehicles, etc.) and use aggregation according to the principle: "buy small and sell big".

Along with digitalization, hard and software steadily improved and new software tools came up. Transparency and traceability is a must in modeling today. Open source modeling is already state-of-the-art and will displace black-box modeling and thinking (see Sect. 5.2). In consequence, new software tools as well as new modeling approaches (such as, e.g., bi-level programming for real-world applications—see Sect. 3.4) can also be applied to the field of modeling smart grid systems.

To further break down the term smart grid system and to derive a systematic taxonomy, smart grid systems can be divided into three sub-groups of application scopes: Small-scale, medium-scale, and large-scale applications. Table 1 classifies a variety of different concepts that are related to smart grid systems in accordance with the scope of application.

Smart home usually denotes residences equipped with smart technologies that enable to provide customers, i.e., residents, with tailored solutions that aim at enhancing the quality of living. Technical devices range from small household devices (e.g., refrigerator, washing machine), sensors, and domestic appliances that monitor and control lightning and heating all the way to integrate decentralized generation and storage applications that optimize the generation and utilization of energy within the smart home. Another strain of smart home solutions deals with energy efficiency and renewable generation integration, where household electricity consumption is optimized to provide electricity grid operators with flexibility in grid management and to avoid congestions in the transportation of electricity in the distribution grid. This can be achieved by direct (real-time) access of grid operators to household information on electricity consumption utilizing smart metering technologies and control of certain devices (e.g., charging of electric vehicles) but can be also controlled via incentive mechanisms. Time-of-use tariffs or dynamic retail tariff structures could provide end-users with real-time electricity price information that incentivize customers to adapt their consumption behavior in an electricity system beneficial way.

[3] A complete listing of all components of smart grids or a definition is out of the scope of this section.

Table 1 Classification of smart grid concepts

Smart grid systems		
Small-scale	Medium-scale	Large-scale
• Smart customer • Smart home • Smart technologies • Smart building • Smart meter	• Quarter solutions • Aggregators • Micro grids	• Smart demand • Renewable integration • Transmission monitoring

Another group of smart grid systems deals with smart applications on a community level. Here, several households in a street of a district or a compound of apartments of a multi-family house (sometimes denoted as quarter) are connected to a network with communication and information systems and optimized as a whole. So-called aggregators represent the group of households which manage the electricity sales and procurement of the entire quarter. The aggregator quite often is usually a service provided by a third-party energy utility. Since it accumulates the energy flows of all quarter members, larger total energy volumes are reached which allows to directly participate at wholesale markets. Again, a necessary prerequisite is that the aggregator is able to gather real-time information of the quarter members' energy consumption and decentralized generation which can be achieved by the installed smart metering systems. Sometimes smart grid systems at a communal level also relate to micro grids. In this context, smart grid systems are understood as an independent subsection of a distribution grid, that can be managed autonomously without being connected to the higher voltage level distribution grid either temporarily in case of fault events or even permanently due to the typology of the supply area, e.g., in very remote areas where geographic restrictions apply. Either way, it supports the grid operator with a higher degree of flexibility for increased integration of decentralized electricity generation and consumption appliances.

Smart grid systems also appear in large-scale applications of energy supply. In various analyses, demand side management (DSM) activities are a predominant characteristic of smart grid systems. DSM also referred to as demand side flexibility or demand response can be defined as the planning, monitoring, and management of activities that stimulate large-scale consumers, such as industrial utilities in changing their consumption behavior which essentially yields into optimized load profiles for system integration of variable generation sources. Again this can be achieved via incentive mechanisms or automated procedures through intelligent technical devices that initiate changes in the electricity consumption, e.g., an adjustment of the production plan. Apart from the generation and demand, smart grid systems can also be part of the transmission of electricity. In this context high-voltage power line monitoring enables optimized utilization of transportation lines. In this way, real-time information on the technical state of the power line (e.g., temperature and power flow) is delivered to the transmission system operator, which can then adjust transportation limits to avoid congestions in the electricity transport. As outlined in the introduction, digitalization is one driver of such emerging power line monitoring systems.

In the following, general model notations of large- and small- to medium-scale, bottom-up demand side modeling as well as bi-level programming will be introduced.

3 Mathematical Notations of Selected Smart Grid Problems

3.1 Large-Scale Models

Against the background of decarbonization as energy-policy target, the expansion of less carbon-intensive technologies related to the conversion, supply, and demand of energy requires the analysis of techno-economic uncertainties. Additionally, to capital-intensive rather central technologies with long lifetime (e.g., power plants or electricity grids), the trend of decentralized energy supply and flexible energy demand at distribution grid level increases the need for assessing positive and negative effects of the system transformation. Large-scale model analyses aim for the evaluation of these complex interactions between different available and future technologies. Besides equilibrium models, which represent the energy sector as part of the whole economy, simulation and optimization models are the main categories for large-scale energy and electricity system modeling. For both methodologies, it is crucial that impact assessment focus on insights from the interrelations between the modeling framework and the techno-economic implementation of the technologies, since models always simplify the real world. Taken this into account, energy and electricity system models enable the evaluation of system configurations to estimate optimal long-term investments and short-term dispatch decisions (optimization models) as well as efficient performances (simulation models) of different energy system components. As mentioned in the introduction, the digitalization and corresponding information and communication technology (ICT) are crucial elements for the decentralized interaction of the relevant components in a smart energy system. However, the modeling of these ICT in large-scale energy models is rather an implicit precondition than explicitly modeled. After giving an overview of central characteristics of large-scale energy system models, further aspects of the inclusion of smart grid technologies in these models will be discussed.

Both modeling methodologies usually apply a techno-economic bottom-up approach. In general, in optimizations models the objective function is formulated as cost minimization to identify least-cost solutions for electricity provision or welfare maximization. For a simplified dispatch model applied for a single node or market (in large-scale models usually an entire country) as illustrated in the following, the objective function minimizes total costs TC as the product of power plant dispatch $G_{i,t}$ and corresponding operational costs oc_i for each technology i and time step t, as shown in Eq. 1. Typically, the time-specific electricity generation costs are composed of plant-specific fuel costs, carbon allowance costs, ramping costs, and further variable costs. These operating costs take technology-specific efficiencies, emission factors, ramp rates as well as availabilities into account. While unit commitment and dispatch models minimize the time-specific electricity generation costs (composed of plant-specific fuel costs, carbon allowance costs, ramping costs, and further variable costs), system planning or investment decision models additionally include investments in relevant generation and storage technologies. Besides operational costs and fixed costs, technology-specific economic parameters like annualized investments are further included in this case.

$$\min_{G_{i,t}} \quad TC = \sum_i \sum_t G_{i,t} \cdot oc_i \tag{1}$$

The objective function is furthermore subjected to a range of restrictions. Thereby, the energy balance represents a key element of dispatch models. As an important constraint (see Eq. 2) in electricity system models, it ensures the balance between electricity generation and electricity demand d_t. Usually, in basic dispatch models the electricity demand is assumed to be inelastic to electricity prices, thus d_t is included as model-exogenous input parameter. For optimization problems, the energy balance as central equation of energy system models implies the variable costs of the dispatch of relevant technologies (including smart grid technologies) to meet the time-dependent electricity demand. Based on this energy balance equation electricity prices can be derived fundamentally. As a basic assumption, the marginal cost and thus, electricity price-driven dispatch of different components of the power system requires direct price signals and respective smart grid infrastructure (in particular information and communication technology (ICT)) for all participants. This is not only true for technologies on the supply side, but also when the demand side becomes more flexible.

$$d_t \; = \; \sum_{i \in I} G_{i,t} \qquad \forall \, t \tag{2}$$

For power plants, minimum restrictions forces the generation to be non-negative, while maximum capacities ensure that electricity generation is not exceeding the installed capacity pc_i. For weather-dependent RES wind and PV, generation time series (usually in hourly resolution) are included as parameters reflecting the weather dependency and the applied feed-in priority.

$$0 \; \leq \; G_{i,t} \; \leq \; pc_i \qquad \forall \, i, t \tag{3}$$

Based on this basic dispatch model, the equations can be extended, particularly when including additional technologies for smart applications and flexibility provision in energy system modeling. Storage provide temporal shifting flexibility to compensate for fluctuations in electricity demand and supply. Regarding a model implementation, storage charging and discharging has to be included in the energy balance, as displayed in Eq. 4. Thereby the set s is a subset of electricity generation technologies ($s \in S \subseteq I$). This allows for restricting the discharging of storages $G_{s,t} (s \in S)$ similarly to power plants. In addition, the charging of storages $P_{s,t}$ has to be restricted to be non-negative as well as to not exceed maximum charging power sp_i as in Eq. 5. Furthermore, the storage energy balance (see Eq. 6) describes the storage inflows and outflows (including a storage efficiency η_s) between two time steps. Thereby, in Eq. 7 the storage level $SL_{s,t}$ is restricted by maximum storage energy capacity sc_i.

$$d_t = \sum_{i \in I} G_{i,t} - \sum_{s \in S} P_{s,t} \quad \forall t \tag{4}$$

$$0 \leq P_{s,t} \leq sp_s \quad \forall s \in S, t \tag{5}$$

$$SL_{s,t} = SL_{s,t-1} - G_{s,t} + P_{s,t} \cdot \eta_s \quad \forall s \in S, t \tag{6}$$

$$0 \leq SL_{s,t} \leq sc_i \quad \forall s \in S, t \tag{7}$$

By increasing the system boundaries of the dispatch model to multiple nodes, the potential to exchange electricity between regions or countries can be included. In large-scale electricity systems the nodes typically represent countries reflecting the import/export of electricity as cross-border flows. This spatial shifting can be seen as crucial flexibility option, particularly in interconnected electricity systems like the European one. When considering numerous countries in dispatch models, an additional set has to be introduced (here c). Furthermore, the flow from one country c to another one is formulated by introducing an alias cc. With export and import, the energy balance of Eq. 2, has to be formulated for each node/country c (see Eq. 8). Additionally, the energy balance between electricity demand, electricity generation as well as storage charging and discharging is extended by imports $EX_{t,cc,c}$ and exports $EX_{t,c,cc}$. Available interconnections between countries are introduced by an adjacent matrix and restricted by existing hourly transfer capacities $ec_{t,c,cc}$ (Eq. 9).

$$\begin{aligned} d_{t,c} &= \sum_{i \in I} G_{i,t,c} - \sum_{s \in S} P_{s,t,c} \\ &+ \sum_{cc \in map(c)} (EX_{t,cc,c} - EX_{t,c,cc}) \quad \forall t, c \end{aligned} \tag{8}$$

$$EX_{t,c,cc} \leq ec_{t,c,cc} \quad \forall c, t \tag{9}$$

The flexibilization of the demand side by demand response (DR) is of high importance in smart grid energy system analysis. To implement demand response measures, Eq. 8 can be further extended by both applications to increase ($LI_{t,c,a}$) and reduce

($LR_{t,c,a}$) electricity load with $a \in A$ as set of demand response (DR) applications, as shown in Eq. 10. The potential of each DR process has to be restricted considering several parameter and characteristics. First, the temporal availability has to be taken into account. While for load reduction the actual electricity demand $dr_{t,c,a}$ of the application forms the upper bound, for load increase the difference between maximum application capacity $drmax_{c,a}$ and current load restricts the availability (see Eqs. 11 and 12). In general, load shifting represents a subcategory of DR, characterized by a balance between the overall load increase and load reduction within a given time frame without influencing the total electricity demand. This is why DR shifting measures can be modeled as storage systems, as in Eq. 13 with $DRL_{t,c,a}$ as virtual storage level. However, since most of the DR processes and appliances have primary purposes (e.g., aluminum production for industry or dish washing for households), the shifting time $tbal_a$ has to be restricted by parameters combining the duration of activation as well as the time in which load reductions and increases must be balanced (Eq. 14). Further restrictions may be included to improve the representation of technical aspect of DR applications. Examples are the limitation of number of activations per day or year (if single processes are implemented). Subsets can be defined to assign applications for load shedding only (load reduction without compensating the load at a later time step), as well as for solely increasing electricity demand. The latter ones are typically introduced as sector coupling technologies (power-to-x) into existing dispatch models, increasing the load due to the electrification of further energy demand sector.

$$\begin{aligned} d_{t,c} &= \sum_{i \in I} G_{i,t,c} - \sum_{s \in S} P_{s,t,c} \\ &+ \sum_{cc \in map(c)} (EX_{t,cc,c} - EX_{t,c,cc}) \\ &+ \sum_{a \in A} (LR_{t,c,a} - LI_{t,c,a}) \quad \forall t, c \end{aligned} \tag{10}$$

$$LR_{t,c,a} \leq dr_{t,c,a} \quad \forall t, c, a \tag{11}$$

$$LI_{t,c,a} \leq drmax_{c,a} - dr_{t,c,a} \quad \forall t, c, a \tag{12}$$

$$DRL_{t,c,a} = DRL_{t-1,c,a} - LR_{t,c,a} + LI_{t,c,a} \quad \forall t, c, a \tag{13}$$

$$\sum_{t}^{t+tbal_a} \left(LR_{t,c,a} - LI_{t,c,a} \right) = 0 \quad \forall t, c, a \tag{14}$$

With the implementation of power-to-X (PtX) technologies, intersections with other energy demand sectors are applied with direct impacts on the energy balance (Eq. 10) in the power system. Larger system boundaries and new actors potentially increase the need for smart communication between the technologies involved. In this sense, a multi-coupled energy system and the corresponding communication of the different sectors can be seen as smart energy system (Ringkjøb et al. 2018). In the model (variations) presented in Eqs. 1–14, particularly the implementation of demand side flexibility represents a simplified application of sector coupling, when flexibility is exploited with PtX technologies like heat pumps, electric vehicles or electrolyzer. These dispatch models still focus on the electricity market but include several options for the electrification of further energy sectors with a respective increase in (time-dependent) electricity demand without a detailed representation of additional energy sectors. Particularly, energy system models enable the comparison of electricity-based energy carriers and the complements of the respective energy sectors. With the expansion of the model system boundaries, further energy end-use sectors, like buildings, industry, and transport have to be included. Accordingly, further mathematical restrictions reflecting additional energy balances and techno-economic constraints have to be considered similarly to the electricity market models.

However, due to computational limitations, the level of detail is generally lower compared to sector-specific models. In general, the bottom-up modeling or simulation implies detailed data input. Therefore, the trade-off between techno-economic detail and computation time is of high relevance regarding large-scale models. Simplifications (i.e., aggregate technologies, countries, time steps) are often necessary to keep the models tractable. However, particularly due to the expansion of fluctuating renewable energies in smart energy system, the temporal resolution is of high importance to represent the increasing variability of the electricity feed-in. Besides the large-scale system perspective, the long-term planning horizon additionally increases the uncertainty of future developments. Especially in optimization models, compromises regarding the technical details are often addressed by applying scenarios and sensitivity analyses.

3.2 Small- to Medium-Scale Models

Models on a smaller spatial scale take on the perspective of a district, quarter, or individual houses. Fundamentally, these models are similar to large-scale models but with important distinctions, explained in the following.

The objective of such a model can be manifold (cf. Mohsenian-Rad and Leon-Garcia 2010; Yu et al. 2013 and Arabali et al. 2012 in Sect. 4.2). For instance, the objective can be the minimization of net costs. A quarter can aim for achieving the most inexpensive way of meeting its electricity demand. Here, c_t is the cost of procuring power, imported via the grid. r_t is remuneration for produced electricity, sold via the grid (Eq. 15). The objective function takes on the perspective of an aggregator which manages the selling and procurement of electricity for an entire quarter.

$$\min \; NC = \sum_t (GRIDIM_t \cdot c_t - GRIDEX_t \cdot r_t) \tag{15}$$

The objective incorporates the goal of avoiding expensive power purchases and can therefore support the grid by reducing peak demand (cf. Mohsenian-Rad and Leon-Garcia 2010; Kahrobaee et al. 2012).

An important distinction to large-scale models is that the energy balance needs to hold for entities, not for market zones or countries. Entities can take on various forms, e.g., individual households. The trade in large-scale models now becomes an exchange with other entities, e.g., with neighbors (Eqs. 16, 17).

$$d_{t,e} = \sum_i (G_{t,i,e} + LR_{t,i,e} - LI_{t,i,e}) + \sum_{ee \in map(e)} (EX_{ee,e} - EX_{e,ee}) \quad \forall t, e \tag{16}$$

Overall, an energy balance needs to be abided by, for the entire quarter. If there is a shortage, including all demand response activities, imports are needed; if generation plus load reduction exceeds the demand, electricity can be exported.

$$d_t = \sum_{i,e} (G_{t,i,e} + LR_{t,i,e} - LI_{t,i,e}) + GRIDIM_t - GRIDEX_t \quad \forall t \tag{17}$$

$$T = \{1, \ldots, 24\}, \; I = \{1, 2, 3\}, \; E = \{1, \ldots, 300\}$$

Noticeably, also the variables for load reduction and increases take on the index t, i and e. This attests to the much higher granularity of small- and medium-scale model. Possibly, for each entity, different technologies can serve as flexible demand (cf. Gottwalt et al. 2016). These models then increase in size, by considering many entities, i.e., generally $|E| > |C|$. Also, the time resolution may increase from hourly to, e.g., quarter-hourly, considering $T = \{1, \ldots, 96\}$ instead (compare Sect. 4.2).

Importantly, generation capacity, demand response, and storage restrictions still apply, along the lines of the large-scale model.

With increasing regionalization the uncertainty in forecasts becomes greater and its consideration gains in importance. This pertains to predicted parameters especially, for instance, demand and availability of renewable energy. For renewable energy it becomes more difficult to accurately predict the availability in a small region. For entire countries, load can often be aggregated by standard load profiles. On a small- or medium scale, the load profiles of smaller entities have to be predicted which is generally more challenging. One method to handle this uncertainty is stochastic optimization (see also Yu et al. 2013).

Consider a quarter which has generation capacities based on renewable energy as well as some flexibility option. The objective of this quarter is the minimization of net costs, comprised of costs for grid imports and negative costs (revenues) for the export to the grid (Eq. 18).

$$\begin{aligned} \min \quad NC = \; & \sum_t \left(GRIDIM_t^{\mathrm{DA}} \cdot c_t^{\mathrm{DA}} - GRIDEX_t^{\mathrm{DA}} \cdot r_t^{\mathrm{DA}} \right) \; + \\ & \sum_{t,s} p_s \cdot \left(GRIDIM_{t,s}^{\mathrm{ID}} \cdot c_{t,s}^{\mathrm{ID}} - GRIDEX_{t,s}^{\mathrm{ID}} \cdot r_{t,s}^{\mathrm{ID}} \right) \end{aligned} \tag{18}$$

The forecasts for demand $d_{t,s}$ and generation availability $cap_{t,i,s}$ are uncertain, i.e., there are a number of possible scenarios s for these parameters. In a first stage, the quarter purchases energy from and sells it to the day-ahead market ($GRIDIM_t^{\mathrm{DA}}$ and $GRIDEX_t^{\mathrm{DA}}$), taking into account the various scenarios. Depending on the realization of scenarios, intraday adjustments have to be made in a second stage ($GRIDIM_{t,s}^{\mathrm{ID}}$ and $GRIDEX_{t,s}^{\mathrm{DA}}$) which, importantly, are scenario-dependent. The scenarios can be weighted by means of p_s, depending on how probable these scenarios are considered to be.

As in the example above, the capacity constraint has to be abided by in each scenario. Also, the energy balance will differ in each scenario with constant elements, such as the day-ahead decisions (from the first stage), as well as scenario-dependent elements like demand, intraday imports/export and load flexibility decisions (Eqs. 19, 20).

$$G_{t,i,e,s} \leq \; cap_{t,i,e,s} \qquad \forall t, i, e, s \tag{19}$$

$$\begin{aligned} d_{t,s} = \; & \sum_{i,e} (G_{t,i,e,s} + LR_{t,i,e,s} - LI_{t,i,e,s}) + \\ & GRIDIM_t^{\mathrm{DA}} - GRIDEX_t^{\mathrm{DA}} + \\ & GRIDIM_{t,s}^{\mathrm{ID}} - GRIDEX_{t,s}^{\mathrm{ID}} \qquad \forall t, s \end{aligned} \tag{20}$$

$$I = \{1, 2, 3\}\,,\; T = \{1, \ldots, 24\}\,,\; E = \{1, \ldots, 300\}\,,\; S = \{1, \ldots, 10\}$$

A key element of stochastic optimization is the consideration of scenarios to account for uncertainties. Questions emerge regarding which and how many scenarios to consider. For instance, scenarios can follow from observed time series or simulated ones, taking into account the key distribution characteristics. To maintain the feasibility of model computations, often scenario reduction techniques are applied (e.g., see Heitsch and Römisch 2003). This becomes especially important with a growing amount of considered entities and a higher time resolution, increasing model complexity. Stochastic programming can, of course, be applied to any model scale. For instance, the effects of intermittent renewable energy feed-in on electricity market are of great interest (Abrell and Kunz 2015).

3.3 Bottom-Up Demand Side Models

The modeling of the future energy demand is of crucial importance for investment and dispatch planning, in particular due to its rising influence on the success for a sustainable energy transition, integration of renewable energy, and smart energy systems. Influencing factors of the energy demand are multiple—for instance, weather and climate conditions, the economic development, technology change as well as changes in policy and consumption behavior. However, several models take only few of these factors into consideration, ensuing incomplete information. Usually in projection models the final energy demand can be carried out for individual energy carriers such as gas, heating oil, district heating, biomass, solar thermal, or electricity (Herbst et al. 2017), whereby in this section the focus is on modeling and projecting electricity demand of individual sectors in a smart grid energy system.

Bottom-up energy demand models derive long-term projections for the future annual energy demand of individual countries based on assumptions on socio-economic data (e.g., gross domestic product, population, evolution of energy carrier prices) and techno-economic data (such as specific consumption, equipment rate, operation time, life time, investment costs). In most bottom-up demand side models the projected annual electricity demand can be further distinguished by year, country, sector, application, and technology (Boßmann et al. 2013). Compared to other sectors, the industrial sector reflects the highest degree of heterogeneity with regard to technologies and energy end-uses. For instance, the industrial sector can be categorized by several sub-sectors such as the iron and steel, non-ferrous metal, paper and printing, chemical, food and drink, tobacco, or engineering industry, to name few. Moreover, a variety of industrial process technologies exist for instance the primary aluminum production, paper production, cement production, electric arc furnace steel, and blast furnace steel production (Herbst et al. 2017). A further distinction can be carried out by defining cross-cutting technologies like lightening or electric motors etc. In contrast, the tertiary sector can be categorized into sub-sectors like trade, hotels, and restaurants, traffic and data transmission, finance, public administration, or health. Those sub-sectors include different energy end-uses such as lightening, electric heating, ventilation, refrigeration and cooling, cooking, data centers, etc. (Herbst et al. 2017). In the residential sector different energy demand groups are distinguished between lightening, sanitary hot water, space heating, among others. Those residential energy demand groups are further categorized in energy end-uses like air conditioning, dish washers, washing machines, dryers, lightening, stoves, computer screens, or television. The energy end-uses can further be classified into technologies and their different efficiency classes (Herbst et al. 2017).

The particularities of each sector such as the granularity, technology structure, and actor heterogeneity as well as the data availability emphasize how complex future energy demand forecasts are. For instance, drivers of the projected energy demand for the tertiary and residential sector are more population-related by, e.g., number of employees and households (Herbst et al. 2017). Whereby, the electricity demand for the sub-sector space cooling in the tertiary sector can be calculated by considering the specific energy demand per m^2 floor area to be cooled and the quantity of the energy service driver, which is the share of cooled floor area per employee. Therefore, the employment in the tertiary sub-sector is an influencing factor which further depends on the development of the gross value added in the sector, demographic trends, and the gross domestic product per capita (Herbst et al. 2017).

Next to the annual electricity demand of a sector, also the hourly electricity load curve is of high importance for designing the future smart grid energy system since the electricity demand and supply needs to be balanced at any time step of the year. The most common and simplified approach in long-term projections is scaling a historical load curve by assuming the future annual electricity demand. This approach implies specific errors as changes in the future electricity load curve are not considered and therefore, the load curve correlates precisely with the historical hourly electricity demand (Boßmann et al. 2013). An alternative approach is the decomposition of the historical load curve by means of sector and application specific electricity loads (cf. Elsland et al. 2013). In this case, the individual load curves are scaled corresponding to the sector and application-discrete annual electricity demand forecast and aggregated to a total sector and/or system load curve.

In the following the disaggregation of the annual electricity demand to the hourly electricity demand for a (smart) energy system is described (based on IAEA 2006): The electricity demand is derived for a given hour (t) of a certain day (d), and period (j, e.g., week, month, etc.) for a specified year by considering the following factors:

1. The average growth rate of the electricity demand over the year (trend).
2. The seasonal variation of electricity demand (e.g., semesters, quarters, months, etc.).
3. The impact of day types (k) in the electricity consumption (i.e., consideration of working days, weekend days).
4. The daily variation of electricity consumption due to certain periods (i.e., morning hours, lunchtime, evening hours, etc.).

These influencing factors are considered by different coefficients which represent the variation of electricity consumption in a sector by referring to the standard load of the sector that is usually calculated for an equivalent working day (IAEA 2006). Before starting to model the annual and hourly electricity demand, the variations in the electricity consumption pattern need to be identified by defining different seasons ($seas$) and day types (k). Therefore, the starting and ending dates of each season during a year need to be defined, similar to the dates of special holiday periods. Further, the sequence of weekdays and representative days (e.g., working days and weekend days) for hourly load variations are to specified.

The coefficient T_j is a correction factor of general trends for the electricity consumption growth during a year. The growth trend coefficient of the gross electricity consumption is calculated on a weekly (j) basis with 52 values for a year (Eq. 21). *GROWTH* is defined as the absolute difference of the annual electricity demand between the current and the reference (past) year related to the total annual electricity demand of the current year.

$$T_j = \quad [1 + GROWTH]^{\left(\frac{i-26}{52}\right)} \quad \forall j \in J = \{1, \ldots, 52\} \tag{21}$$

The seasonal coefficient K_j considers the impact of different seasons on the electricity consumption pattern in a sector for a certain time period (j) which can be a semester, a quarter, a month or a week. Assuming weeks as specified time period, K_j can be defined as the average weekly weight of the yearly electricity consumption. The sum of the coefficients of a year equals the value of the total number of periods into which the year is divided (e.g., 52 weeks).

The daily ponderation coefficient $P_{j,d}$ considers the fluctuations within the electricity consumption pattern of different day types, i.e., working days, Saturdays and Sundays. Therefore, the electricity demand of every time step of the reference (past) year is compared to the electricity demand of an equivalent working day, which has the relative weight of 1. At first, the yearly average electricity demand of each working day needs to be calculated. The value which is nearest to the mean of all working days can be considered as equivalent working day. The other day types are weighted to their relative electricity demand compared to the equivalent working day (e.g., Saturday might achieve a 0.8 of a working day, etc.). The coefficient $P_{j,d}$ varies over the year depending on the defined seasonal periods (j) and the day types.

Subsequently, the average electricity demand of an equivalent working day can be calculated dividing the total annual electricity demand of the current year by the total number of equivalent working days, which is the sum product of the trend coefficient T_j, the seasonal coefficient K_j and the daily ponderation coefficient $P_{j,d}$ over all calendar days cd of a year (Eq. 22).

$$D_{cd} = D \quad / \sum_{j,d \in map(cd)} \left(T_j \cdot K_j {}^* P_{j,d}\right), \tag{22}$$

$$\forall cd \in CD = \{1, \ldots, 365\}, \, j \in J = \{1, \ldots, 52\}, \, d \in D = \{1, \ldots, 7\}.$$

To disaggregate the average electricity demand of an equivalent working day D_{wd}, the hourly load coefficient $LC_{t,d}$ is needed. $LC_{t,d}$ reflects the weighted hourly electricity demand over 24 h of a day. The coefficient is calculated by the hourly electricity demand related to the daily electricity demand of the reference year multiplied by 24 h. The sum of all coefficients of a day is equal to 24. Consequently, the hourly electricity demand can be calculated by multiplying the average electricity demand of an equivalent working day with the hourly coefficient divided by 24 h (Eq. 23).

$$D_t = D_{cd} \cdot LC_{t,d}/24 \quad \forall t \in T = \{1, \ldots, 8760\}, d \in D = \{1, \ldots, 7\}. \tag{23}$$

In context of modeling smart grids, the electricity demand side needs to be flexible to balance the electricity supply at any time (e.g., facilitating the integration of high shares of intermittent). As described in Sects. 3.1 and 3.2, the flexibilization of the electricity demand side can be achieved by integrating demand response applications ($a \in A$) which can increase ($LI_{t,a}$) and reduce ($LR_{t,a}$) electricity load in certain time periods when needed. The potentials of DR applications are constrained by several parameters and characteristics as described before in Sect. 3.1 and in Eqs. 11–14.

3.4 Bi-level Programs

The above formulated large-, medium- and small-scale models represent typical decision problems of individual energy stakeholders in a smart grid environment that can be modeled as a single linear or non-linear optimization problem. While these models are useful approaches to investigate the various interactions in smart grid systems, they sometimes fall short of representing the heterogeneity in the preferences of multiple stakeholders and the potentially involved interaction between them. Forms of interactions between sub-systems of a smart grid can create a stronger coupling between the demand- and supply side. For example the participation of end-users which own distributed flexible energy applications, e.g., electric vehicles or heat pumps, in corresponding electricity markets. Whenever sub-systems of a smart grid are managed by individual stakeholders, the model-based representation entails the inclusion of different objectives in the optimization problem, which requires other modeling techniques than those discussed above.

Let us assume the following example: Consider an aggregator, which buys energy at the wholesale markets to serve a given load of a residential quarter. The aggregator can decide on a dynamic tariff scheme offered to the customers representing the members of the quarter. Since wholesale prices vary over the course of a day the aggregator tries to design a tariff scheme that incentivizes the customers to shift their energy consumption into times with low wholesale prices. Aggregators usually try to maximize their profits, which can be calculated by subtracting the energy procurement cost at the wholesale market from the revenues from electricity consumption of the quarter. An intelligent communication infrastructure allows the aggregator to have full information on the customer's demand profile.

Furthermore, members of the quarter adjust their electricity consumption behavior in a way that minimizes their total expenses for electricity supply, which is strongly affected by the provided electricity tariff. However, usually only a fraction of the quarter's electricity demand can be considered flexible as specific technical limitations of the household devices prevent a full flexible operation.

As we can see here, the optimal decision on the design of the tariff scheme in the optimization problem of the aggregator is constrained to be optimal solutions to the optimization problem of the members of the living quarter. Such a framework refers to the class of bi-level optimization programs, in which one optimization problem is nested into another. In the outlined example the lower level optimization problem of the quarter members is included in the upper level optimization problem of the aggregator. Since the upper level variables are considered as fixed parameters and not decision variables in the lower level problem, i.e., the provided dynamic electricity tariff cannot be actively controlled by the quarter members, the optimization problem follows a hierarchical structure as in the well-known Stackelberg leader/follower game.

It is important to understand that the raised example is just one possible application of a bi-level program in the smart grid context. More generally, the upper level can be considered as a strategic decision-maker, who anticipates a feed-back from the lower level problem. Strategic decisions could be manifold, e.g., investment decisions, tariff-design or other regulatory questions. The lower level problem could be a response from the energy management of a single household, an integrated residential quarter, or an entire electricity market. As a smart grid typically exhibits a strong interaction between different sub-systems of an energy system, bi-level programming can be seen as a powerful tool for its model-based investigation.

In the generalized mathematical form we can write a bi-level program as follows (Bylling 2018):

$$\min \mathrm{f}_1(\mathrm{x}, \mathrm{y}^*) \tag{24}$$

$$\text{s.t. } \mathrm{g}_1(\mathrm{x}, \mathrm{y}^*) \leq 0 \tag{25}$$

$$\mathrm{h}_1(\mathrm{x}, \mathrm{y}^*) = 0 \tag{26}$$

$$\mathrm{y}^* \in \operatorname{argmin} \{\mathrm{f}_2(\mathrm{x}, \mathrm{y}) \tag{27}$$

$$\text{s.t.} \quad \mathrm{g}_2(\mathrm{x}, \mathrm{y}) \leq 0 \tag{28}$$

$$\mathrm{h}_2(\mathrm{x}, \mathrm{y}) = 0\}. \tag{29}$$

In this formulation, f_1 represents the objective function of the upper-level with the decision vector x (Eq. 24) and f_2 the objective of the lower level with the decision vector y (Eq. 27). Likewise, we have two sets of constraints represented by g_1 and h_1 (Eqs. 25, 26) as well as g_2 and h_2 (Eqs. 28, 29). The upper level decides on the values of x which minimize the objective f_1 anticipating the response of the lower level summarized in the values of the decision vector y, which minimizes the objective f_2. Due to the nested structure, solving bi-level programs is sometimes a challenge. Existing solution methods involve the replacement of the lower level problem by their necessary and sufficient Karush-Kuhn-Tucker (KKT) system following complementarity theory. The KKT system represents optimality conditions that must hold in the optimal solution of the lower level problem and which again can be reformulated as a set of additional inequality constraints. The newly derived set of constraints can then be taken together with the upper-level problem, recasting a single optimization problem that can be solved with commercial solvers for linear and non-linear programs.

4 Overview of Existing Modeling Approaches of Smart Grid Systems

The following subsections provide an overview of existing modeling approaches addressing smart grid components in large-scale energy models with a system perspective. Secondly, an overview of existing small- to medium-scale models narrowing down the scope on consumers' perspective is provided. Thirdly, to further increase the detail of the energy demand side modeling, bottom-up demand side models are examined. In general, the distinction between "traditional" energy models and smart energy models is fluid, and a strict differentiation is not possible.

4.1 Literature on Large-Scale Models

Among others, model descriptions for large-scale energy system models can be found in Panos and Lehtilä (2016) or Hidalgo Gonzalez et al. (2014). Examples for investment models can be found in Gils et al. (2017) or Zerrahn and Schill (2015). In addition to operational costs and fixed costs as well as technology-specific economic and technological parameter, relevant energy policies such as feed-in priorities of (weather-dependent) renewable energies are usually employed with country-specific time-dependent feed-in time series (e.g., in Child et al. 2019; Nitsch et al. 2012). The EU emissions trading scheme (ETS) can be represented by explicitly implementing a carbon cap or budget constraining the total amount of emissions allowed, as in Hobbie et al. (2019), Capros et al. (2016), Möst and Keles (2010) or by implicitly specifying prices for emission allowances, as for example in Zöphel et al. (2019) and Oei et al. (2014).

Generally, in a smart grid system with different levels of (de-) decentralization in supply and demand balancing, combinations of flexibility options are object of large-scale energy system analysis. Various technologies for supplying electricity (e.g., power plants), shifting energy (e.g., storages or transmission grids) or increasing electricity load (i.e., power-to-X (PtX), such as heat pumps, electrolyzer for hydrogen production or battery electric vehicles) are analyzed with different scopes. While for example Cebulla and Fichter (2017) analyze the influence of different renewable energy sources (RES) shares on investments in power plants and storages in a regional case study, Brijs et al. (2017) apply their analysis in a similar set-up for Belgium. In contrast, examples for an increase in system boundaries can be found in Connolly et al. (2016) or Koch et al. (2015), where additionally investments in transmission grid expansions, demand side management (DSM) and PtX technologies are analyzed for Europe and different shares of RES. These models still focus on the electricity market but include several options for the electrification of further energy sectors (sector coupling) with a respective increase in (time-dependent) electricity demand without a detailed representation of additional energy sectors. As an example, Lund and Kempton (2008) simulate the interaction between different electric vehicle charging strategies and levels of wind expansion on the amount of RES-based excess generation on national level. In general, the majority of optimization problems for electricity markets and energy system is a linear or mixed integer program. Nevertheless, non-linear formulations exist as well (e.g., when optimizing the number of power plants or including price elastic demand). An overview is given for example in Fernández-Blanco Carramolino et al. (2017) or Möst and Keles (2010). Further examples for smart grid applications are methods on the electricity supply side like real-time RES generation (as in Bottaccioli et al. 2017) or real-time pricing to balance electricity demand (as in Tao and Gao 2020). Furthermore, an emerging research field and application for simulations models is the estimation of time of arrival for vehicle-to-grid measures as in Luo et al. (2016). A combination of optimization and simulation models is often applied to derive optimal scenarios also for mid-term and long-term time horizons as well as to evaluate the efficiency of these scenarios in simulations with higher temporal resolution on shorter time scale. An example can be found in Rosen (2007), where optimization and simulation approaches are used to capture both, long-term changes and variability of renewable integration.

In a large-scale electricity market perspectives, the aspect of decentralization, introduced in Sect. 1, is often assessed by comparing scenario frameworks including technologies generally assessed as decentral, as for example photovoltaics (PV) rooftop system and decentral heat pumps, with more central systems characterized by a higher share of for example wind offshore farms and combined heat and power (CHP) plants. These kinds of modeling approaches are described for example in Zöphel et al. (2019) with a European system perspective. Furthermore, with large-scale energy system models selected decentral or central approaches can be analyzed. On the one side, both simulation and optimization models are applied in the literature to evaluate possible system effects of scaling-up small-scale smart applications and examine the large-scale impacts of an expansion of decentral smart technologies. While, as already mentioned, the application of optimization models tends to assume the presence of smart grid technology, simulations often focus on the efficiency of suitable algorithms and control strategies for the communications between the components involved. In Bazan et al. (2015), a smart grid simulation including an explicitly simulated controller for single houses is up-scaled and applied for 200,000 households to evaluate the influence of battery and PV size on the average electricity costs in different system configurations. Similarly, in Schill et al. (2017) an optimization model is used to analyze the interactions between different prosumage strategies for PV-battery systems with varying levels of self-consumption and optimal storage investments in Germany. Thereby, the optimal coverage of the prosumer electricity demand is modeled as minimum restriction, while the realization of the self-consumption is a result of the optimization. On the other side, rather central smart grid applications address the optimization of grid operation, usage, and infrastructure as well as integrating large-scale intermittent generation. As an example, Hinz (2017) discusses the impact of decentralization and RES expansion in the German electricity system on the provision of voltage stability and reactive power management in a combination of non-linear and a linearized techno-economic grid models.

Within this range of model applications and system boundaries, the flexibilization of the energy demand side in general (see for example in Müller and Möst (2018) and Ladwig (2018)) and sector coupling in particular with corresponding ICT and the resulting bi-directional power flows, as in Mathiesen et al. (2015), are in the focus of smart grid model analyses.

4.2 Literature on Small- to Medium-Scale Models

Conejo et al. (2010) state that most existing literature assumes a "consumer sufficiently large to participate in the electricity market to minimize its energy procurement costs." Due to the small and distributed nature of end consumers, individual households are often considered in an aggregated way. Nevertheless, this necessitates the consideration of consumers' or prosumers' involvement and behavior as active participants in energy systems and markets, coming along with smart grids and new contractual agreements, such as real-time pricing, resulting in new decision-making models. This brings forth new challenges as described in Sect. 1.

Optimization in smart grids from a consumer perspective often involves the formulation of a cost minimization or utility maximization problem. The objective function typically is the reduction of the end-users' electricity payments. For this, residential load control schemes are designed and deployed, deciding on energy consumption in the household. Because of the manifold objectives of end consumers, optimization tasks can deviate from purely technical and cost-minimizing perspectives and involve factors regarding customers' satisfaction and comfort. Further objectives can include, for instance, a trade-off between electricity bill and waiting time for operation, as in Mohsenian-Rad and Leon-Garcia (2010) or similarly a trade-off between costs and quality of services in Yu et al. (2013), with a minimization in consumer dissatisfaction, measured by temperature deviations. From a system perspective, more technical objectives can be accounted for, such as Arabali et al. (2012), who evaluate a compromise between risk of failure to meet demand and generation cost for different levels of wind and PV. Lastly, with growing shares of decentralized energy generation, individual consumers or energy communities can pursue targets for certain levels of autarky or also complete self-sufficiency. As described in the introduction, this presents new challenges and model objectives beside cost-minimizing considerations.

While most load management approaches take on the grid's perspective, with smart grids, "bi-directional data flow and interoperability between homes and the grid" (Kahrobaee et al. 2012) cause the possibility to optimize individual consumption. Existing literature repeatedly indicates that these cost-reducing schemes on an individual basis often simultaneously create benefits for utility companies and the operation of the grid due to the reduction of peak-to-average load ratios, e.g., Mohsenian-Rad and Leon-Garcia (2010), Kahrobaee et al. (2012). The maximization of individual utilities through cost-saving incentives, brought forth by pricing signals and technical possibilities for flexible demand, usually also leads to overall welfare benefits. This goes along with the grid infrastructure-related questions mentioned in Sect. 1, which arise with a growing expansion of smart grids, namely adequate transmission capacities to properly integrate consumers and steering their behavior to be beneficial for the grid.

As mentioned before, demand has to become increasingly flexible with growing utilization of renewable energy sources. Thus, cost-reducing decisions not only involve demand response, as e.g., modeled in Conejo et al. (2010), but additionally the generation and storage of electricity. Kahrobaee et al. (2012) mention that models in earlier literature have not fully utilized smart home features, they mention the oversimplification and overly restricted model in Pipattanasomporn et al. (2009), the inability to generate power in Ramchurn et al. (2011) or the lack of demand response in Vytelingum et al. (2010). When all possible consumer decisions are taken into account in the modeling of smart grids and their separate entities, passive load curves do not suffice due to the active participation of end-users or smart homes. Beside these modeling aspects, questions emerge regarding the "proper" storage technologies to balance renewable energy feed-in, which are characterized by technical considerations, e.g., storage capacity and discharge time, and feasibility requirements.

Often, linear optimization is used, as in Conejo et al. (2010) and Mohsenian-Rad and Leon-Garcia (2010). Yu et al. (2013) use a mixed integer multi-time scale stochastic optimization to model a home energy management that controls energy consumption in response to dynamic pricing. Studies have increasingly used multiagent-system-based approaches, e.g., Kahrobaee et al. (2012), in which smart homes are agents in a smart grid environment that can consume, generate and store electricity, making autonomous decisions to manage these components while interacting with the grid. The objective is to minimize the cost of electricity. As pointed out in Sect. 1, agent-based models will likely gain in importance. Kahrobaee et al. (2012) describe that multiagent systems are advantageous due to their versatility and scalability. Furthermore, they are able to model stochastic and dynamic interactions among agents, i.e., end-users or homes, and between homes and the grid. With multiagent models, transition periods in the simulation result in an equilibrium "as an emergent behavior of the agents".

The time resolution of smart grid models is often hourly (e.g., Mohsenian-Rad and Leon-Garcia 2010; Gottwalt et al. 2016). Case studies range from covering one day up to several months. The geographical scope is similarly diverse, ranging from single households to an accumulation of hundreds or thousands of consumers. This attests to the vastly different perspectives of smart grid modeling, ranging from micro/household to macro/community perspectives.

One strand of literature considers single households or consumers, e.g., Conejo et al. (2010), Mohsenian-Rad and Leon-Garcia (2010), Yu et al. (2013) and Adika and Wang (2013). More comprehensive approaches consider numerous households with different household devices, which directly respond to different load and renewable energy generation. Stadler et al. (2009), electric vehicles in Schuller et al. (2015), heating/cooling systems in Hakimi and Moghaddas-Tafreshi (2014) or control approaches for stationary batteries in van de Ven et al. (2013). Considering numerous devices necessitates a categorization regarding the technical possibilities for demand flexibility. For instance, Gottwalt et al. (2016) differentiate between automatically controlled devices, such as refrigerators and storages for water heaters as well as semi-automatic devices like dishwashers and washing machines, which require previous user interactions.

When numerous households are considered, their participation can be modeled as aggregators, e.g., Ottesen et al. (2016) and Iria et al. (2018). The aggregator can control the prosumers' flexible energy units. Ottesen et al. (2016) model a two-stage stochastic mixed integer linear program with bidding decisions in the first stage and scheduling in the second. The case study with a diverse portfolio of prosumers attests to the heterogeneity of consumers and behavior. Iria et al. (2018) model an aggregator of small prosumers in the energy and tertiary reserve markets by means of a two-stage scenario-based stochastic optimization model. It becomes evident that a multitude of uncertainties have to be considered, such as renewable power generation, electricity demand, outdoor temperature, end-users' behavior, and preferences. Results include that system flexibility increases with an aggregator (Ottesen et al. 2016) and the reduction of bidding net costs under the consideration of flexible strategies (Iria et al. 2018).

A further example for the modeling of aggregators includes Gottwalt et al. (2016), who investigate the interactions between different shares of renewable energy and the utilization of demand side flexibility. For this, the authors provide a comprehensive centralized scheduling model to make use of demand flexibility in a residential micro-grid. An aggregator with full information dispatches controllable devices with the objective of cost minimization, considering power system balances and device constraints. The problem is formulated as a mixed integer linear program. The analysis derives cost reduction potentials of flexible loads and recommendations for electric utilities to structure their renewable portfolio. Recommendations include that aggregators should incentivize customers to own the according appliances depending on the renewable mix. This attests to the importance of interactions between the aggregated level and individual consumer behavior.

Besides the challenge of modeling smart grids and their participants, existing studies also point out the lack of opportunities and incentives for small consumers to participate in the market, e.g., Zepter et al. (2019). This suggests that beside the consideration of technical constraints and cost-minimizing approaches, consumer involvement poses another challenge. For instance, Mohsenian-Rad and Leon-Garcia (2010) argue that the two major barriers for the full potential utilization of real-time pricing are the lacking knowledge of consumers regarding the response to time-varying prices and effective building automation systems.

Zepter et al. (2019) highlight that the challenge of integrating local markets into the wholesale market has not been sufficiently addressed. They propose a framework, the Smart elecTricity Exchange Platform (STEP), which involves the coordination of operation of supply-demand decisions and provides an interface between wholesale electricity markets and prosumer communities. Rather than considering aggregated distributed energy generation, as is the case in most existing literature, Zepter et al.

(2019) take into account the local distribution and peer-to-peer trading, which enables households to balance out deviations from the community's day-ahead market commitment in the intraday market. This addresses the above-mentioned question regarding the interlinkage of markets to contribute to the integration of renewable energy sources. The authors state that approaches like their proposed one bear great cost reduction potentials. The study attests to multitude of modeling aspects, which need to be addressed to adequately map the different options that end-users can utilize in (future) smart grids.

The modeling of smart grids on a small- to medium-scale encompasses a wide array of approaches and perspectives. When individual households are considered, pure cost-oriented objectives do not capture all facets of consumer behavior. While demand response as well as the decentralized generation and storage of energy are subject to technical constraints, consumers likely have expectations regarding their comfort and convenience level. Aggregators can help to bundle, coordinate and market the potential of communities. Utilizing the full potential of prosumers and smart grids necessitates the technical implementation as well as the provision of information to end consumers.

4.3 Literature on Bottom-Up Demand Side Models

With the crucial role of energy end-users in smart grid systems, energy demand models are of high importance as forecasting energy demand and supply is essential for ensuring a reliable and secure energy system. Furthermore, demand side models allow analyses in the context of decarbonization, decentralization, and digitalization (cf. Sect. 2).

In a system perspective, the annual energy demand is strongly related to energy prices, the gross domestic product (GDP), and population growth (Suganthi and Samuel 2012). The electricity demand is influenced by technological and socio-economic drivers, such as economic growth, energy efficiency, urbanization, per capita income, support schemes for renewable energy sources and other low-carbon energy carriers, as well as by electrification and technological progress in electricity generation technologies. To measure the impact of intermittent RES and to estimate operational flexibility of the future power system, the need arises to model and forecast electricity demand in high resolution (Adeoye and Spataru 2019). National hourly electricity demand pursue a periodic and predictable daily pattern. Changes in the pattern of electricity demand depend mainly on energy efficiency improvements, de-industrialization, and the increasing electrification of the industry, heat, and transport sector (Boßmann and Staffell 2015).

However, most recent studies forecast the future electricity demand by neglecting the further development of the electricity consumption pattern. The studies assume simplified that the hourly electricity load curve maintains its shape by scaling up equally in all hours. Consequently, capacity requirements for flexibility options and peak load technologies increase, and full load hours as well as the profitability of conventional base-load and mid-load power plants decrease (Boßmann and Staffell 2015). Hence, significant transformations of the electricity load curve can evolve in future with the diffusion of new and the phase-out of existing technologies. The changes can have substantial effects, which are crucial to integrate in the modeling of smart grid systems. For instance, the need to cover greater residual load peaks could arise which can be balanced with storages, DSM applications, interconnections, and peak load capacities or the need to flatten hours with negative residual load by curtailing the excess of renewable energy sources (Boßmann and Staffell 2015).

Several studies have modeled and projected the hourly electricity demand and the annual electricity demand by aggregating the individual sectors on country, regional, and sector level. Typical sectors are the industrial, tertiary, residential, and transport sector. Different methodologies to forecast electricity demand on annual or hourly consumption exist. Suganthi and Samuel (2012) have conducted a literature review that provides a comprehensive overview of energy demand forecasting techniques.

The following section focus on bottom-up energy demand models as these models are common for modeling smart grid energy systems. Bottom-up energy demand models are characterized by their high degree of technological detail, which allows to model several, and clearly defined technologies to assess future energy demand and supply (Fleiter et al. 2011). As already mentioned, due to its technological accuracy and explicitness, bottom-up models are applied to model effects of sector- or technology-oriented policies (Gillingham et al. 2008).

Different mathematical formulations of bottom-up models have been developed that can be categorized in partial equilibrium models, optimization models, simulation models, and multiagent models (Herbst et al. 2012). The development of energy end-uses and their respective energy efficiencies estimates the future energy demand. Within all bottom-up energy demand models the demand forecasts are directly linked with the technological structure of the energy system (Fleiter et al. 2011). In the following, bottom-up energy demand models are classified in annual energy demand (Sect. 4.3.1) and in hourly electricity demand projection models (Sect. 4.3.2).

4.3.1 Bottom-Up Modeling of Annual Energy Demand

New challenges have to be faced by policymakers, therefore annual energy forecasting models are applied to assess the potential impact of new policies and to support the decision-making process (Worrell et al. 2004). In general, though optimization and partial equilibrium models are applied as well, the majority of the approaches are simulations. There exist various scopes for both, the geographical coverage ranging from national to international level as well as to system boundaries. Regarding the latter one, analyses can focus on a single sectors, like the industry sector, or on multi-coupled sectors including different energy end-use sectors.

Bottom-up energy demand models are applied to support scenario designs and analyses for the long-term evolution of energy demand and GHG emissions in different sectoral and geographical scales. Annual energy demand models aim to integrate policies and changes in the socio-economic framework including the consideration of a broad range of greenhouse gases (GHG) mitigation options with high degree of technological detail. Usually the three sectors—industrial, tertiary, and residential sector—are depicted by those models and characterized by different specific data requirements, e.g., the production in the industry sector, number of employees in the tertiary sector or number of households in the residential sector. Further input parameters are the main drivers as gross domestic product, populations growth, energy prices by energy carrier, temperature (heating and cooling degree days), and business cycles. Furthermore, price-based policies are considered as taxes, CO_2 prices, market-based instruments (e.g., the EU emissions trading scheme (ETS)), subsidies as well as operational expenditures. Additionally, structural information as the energy balance, CO_2 balance, and the technology distribution as well as technology parameters including behavioral assumptions are reflected (Herbst et al. 2017). The outcome of annual energy demand models can be disaggregated in high resolution from sectors (e.g., residential) and sub-sectors (e.g., industrial combined heat and power (CHP)), as well as energy end-uses (e.g., space heating), technologies like industrial CHP and energy carriers (e.g., natural gas). The diffusion of technologies is the result of individual investment decisions over a specific time period to cover the energy demand in different sectors. Therefore, the investment decisions are commonly modeled as discrete choice process, where companies and households have miscellaneous technology choices to satisfy a specific energy demand (Herbst et al. 2017). This is typically implemented as logit approach considering the total cost of ownership (TCO). Thus, the simulation algorithm considers market heterogeneity and non-rational behavior that leads to price sensitive technology and energy demand developments (Herbst et al. 2017).

In the recent past, especially due to decarbonization targets for all sectors, the importance for building park models has increased. Building park models are bottom-up demand side models that simulate the development of energy-related equipment and the resulting energy demand in the building sector. For this purpose, building technology, construction engineering, energy-specific, and economic parameters, such as investments and life cycle costs as well as influencing factors as energy end-user prices, interest rates, energy and emissions taxes as well as subsidies, are considered. Building park models show future costs and technology developments in the field of energy efficiency and for the use and provision of decentralized heating, cooling, and electricity (TEP 2020). Furthermore, demand side models for the building sector are used to define energy and climate targets, for impact analyses and evaluations of energy climate and policy measures (ex-ante and ex-post), strategic and operational energy planning, urban planning, network expansion planning, and evaluation of network renewal projects (electricity, gas, heating, cooling, etc.). Additionally, the models support the creation of emission and energy statistics, material flow analyses as well as the management of building portfolios or the conduction of market studies. Therefore, past and future changes in national, regional, urban or municipal building parks are simulated over several decades to receive evaluation indicators such as the electricity and energy demand for energy sources, maximum load (electricity, heat), primary energy consumption, carbon and GHG or material flows (new buildings, existing buildings, dismantling). Moreover, the building specific modeling provides insights concerning the optimal choice of heating systems for new buildings and renovations, or of repairing vs. energy-efficient renovation. Furthermore, results as the efficiency level of renovations, devices, and building technology components can be gained.

4.3.2 Bottom-Up Modeling of Hourly Electricity Demand

As stated before, most studies apply a simplified scaling approach of a historical load curve corresponding to an annual demand forecast to assess the future load curve. Within the scope of long-term energy system modeling more sophisticated approaches are required as the diffusion of new technologies and the phase-out of existing technologies may lead to significant changes within the pattern of the hourly future load curve (Boßmann and Staffell 2015). The majority of the studies assess load curve projections for single sectors or consumers (Voulis et al. 2017; Hayn et al. 2018; Lee et al. 2019). Other studies focus on regional load curve projections in a specific country (Riva et al. 2019; Boßmann and Staffell 2015). Further literature analyzes specific characteristics of the hourly future load curve as the hourly peak electricity demand (Hainoun 2009) or the load duration curve (Poulin et al. 2008).

Most bottom-up models assess the hourly electricity demand in the residential sector for an entire year. Hourly electricity load curves for households have been modeled for the United States (Capasso et al. 1994), India (Riva et al. 2019), United Kingdom (Richardson et al. 2010), and Finland (Paatero and Lund 2006) to name few. These models integrate behavioral, social, technical, and economic data, as well as weather data to model the electricity demand for representative households and its consumers (Adeoye and Spataru 2019). Typically, these are simulation models taking for instance the increasing diffusion of e-mobility and further decentralized electricity generation appliances into account. Commonly, electricity load curves for specific household appliances (e.g., heating technologies) or the electricity consumption by electric vehicles in households are estimated. For instance, the hourly load curve of e-mobility in the residential sector can be derived from the number of battery electric vehicle (BEV) and plug-in hybrid electric vehicle (PHEV), the electric driving share of plug-in hybrid electric vehicle (PHEV), the traveled kilometers per year, the energy consumption of battery electric vehicle (BEV) and PHEV, multiplied by the share of charged electricity (Elsland et al. 2013). The future residential electricity demand is characterized by increasing volatility due to demand shifts from night-time to day-time hours caused by a growing number of information and communication technologies, while electro-mobility increases evening demand peaks. Electricity generation by photovoltaic can compensate the additional demand due to electric vehicles, if the decentralized electricity generation can encounter the electricity demand of demand side management applications and storage systems.

In contrast to the modeling of the residential sector, there exist only few studies and bottom-up models that focus on the hourly electric load forecasting for all sectors (industry, tertiary, residential, and transport sector). For instance, Pina et al. (2011), Hainoun (2009), or Boßmann and Staffell (2015) assess user specific load profiles of representative customers to composite the future load curve within the industry, tertiary, and residential sector, based on empirical data. Therefore, the profiles are scaled to the annual electricity demand forecast and aggregated to estimate the entire hourly electricity load curve. Hourly electric load forecast models are applied to assess the future pattern of the electricity system load curve at national level for the long-term (e.g., until 2050) by considering all demand side sectors. In general, the projection of hourly electricity load curves can be realized by the deformation of the load curve due to structural changes on the demand side and due to the diffusion of new appliances (e.g., e-mobility) by applying a partial decomposition approach (cf. Boßmann and Staffell 2015). The annual electricity demand projection is an exogenous model input and is necessary to identify significant electricity consumption increases or decreases by relevant appliances over the long-term perspective (Zöphel et al. 2019). Appliance specific load profiles from surveys, official databases, or simulation models are used, to generate load curves for all appliances, according to the annual demand in the base year. Consequently, the specific appliance load curves and the remaining load curve are scaled for all projection years with regard to the electricity demand evolution. Further, the load curve can be adjusted by the flexible dispatch of DSM applications. The DSM appliances' load is based on day-ahead price signals, and scheduled from hours with high prices to hours with low prices. With this approach the least-cost dispatch of DSM appliances from a consumer perspective can be estimated in order to smooth the residual load (Boßmann and Staffell 2015).

Finally, forecasting hourly electricity load curves across different sectors is of crucial importance for modeling smart grid systems, since the future electricity demand will transform significantly as the diffusion of new and the phase-out of existing technologies have a great impact on the daily electricity consumption pattern.

5 New and Other Modeling Trends

5.1 Forecasting: High Resolution of Weather Data and Time Series

The combination of more decentralized power generation and more active consumer behavior, which arises from the proliferation of prosumers, also contributes to new challenges of forecasting tasks. In a smart energy system with high shares of weather-dependent renewable energies, especially weather forecasts are gaining strongly in significance. Da Silva et al. (2013) describe that the transition toward an information-driven smart grid as well as local electricity markets depends on accurate forecasts of its participants' demand and generation. Predictions for renewable energy generation at a higher geographical resolution will become increasingly important when planning and operating local energy systems and communities. The analysis by Schönheit and Möst (2019) describes the different distributions of day-ahead prediction errors for wind speeds. The authors find that the error distributions differ across Germany, which affects the uncertainty connected with local availability of wind energy. This highlights the necessity of accurate prediction techniques on a small-scale level.

Sobri et al. (2018) and Das et al. (2018), which both provide overviews of photovoltaic power forecasting techniques, point out that PV is an often-used technology, also for "stand-alone" or "off-grid" networks. This pertains to small grid-connected consumers as well due to the integration of PV in the buildings of prosumers. Lorenz et al. (2012) describe that on a local level, smart grid applications result in an increased need for PV power forecasting. The authors propose an approach for regional PV power, specifically focusing on snow detection. Das et al. (2018) describe that PV capacities have grown substantially on the past year, but their effect on the grid necessitates accurate forecasting techniques to maintain stability and reliability and aid the modeling and planning of solar photovoltaic plants. To meet the complex task of taking into account the weather dependency when predicting solar energy generation, often neural network-based approaches are used, as in Rodriguez et al. (2018). Shang and Wei (2018) deploy support vector forecast solar power output.

Additionally, with rising participation of consumers in the electricity markets, electricity price predictions may also gain in importance at household level. In general, electricity price predictions are already at high importance since liberalization. Wang et al. (2019) describe that day-ahead electricity price forecasting is an important element for decision-making of market participants. This includes consumers in a market-oriented environment as stated by Zhang et al. (2019). Neural networks are also applied for price predictions, e.g., in Kuo and Huang (2018) or Chow et al. (2012). Wang et al. (2019) use a weighted voting mechanism to combine numerous predictions and achieve better performance than with unified modeling. Forecasting combination, i.e., taking the (weighted) average of multiple forecasts, is also used by Ziel and Weron (2018).

Finally, forecasts are not only needed for generation and prices but also demand. When demand is considered in models on the level of households or communities, e.g., as opposed to high-voltage grid nodes, a higher geographical resolution is necessary for forecasting models. Yu et al. (2015) state that energy resource management in smart grids face the challenge of fluctuations, both on the demand and the supply side. They deploy several machine learning-based approaches and neural networks to forecast energy usage. Goude et al. (2013) state that innovative technologies, such as smart grids, create challenges for electric load forecasting. They propose a semi-parametric approach based on generalized additive models theory to predict electric load at substations. Da Silva et al. (2013) tackle the challenge of forecasting individual demand by the creation of groups. They also show that groups can act as a single unit on the market and use the positive effects of aggregation on forecasting accuracy.

5.2 *Open Source, Transparency, and New Software Tools*

As a general shift toward open methods can be recognized in the energy system modeling community, this development has also influenced research in the context of smart grid modeling. Open methods generally refer to the disclosure of associated source codes, datasets, and documentation with appropriate licensing for reuse, modification, and republication of modeling works. Open modeling is often accompanied by the development and maintenance of open power system data bases that share modeling data among researchers. As summarized in Morrison (2018) main drivers that can explain this shift in modeling paradigms include the desire for improved public transparency, the need for scientific reproducibility and the believe that open methods potentially improve academic productivity and quality.

While reproducibility of research results and general research performance strongly relates to academia, a high degree of transparency is also of great importance in real-world smart grid applications. With a strong involvement of energy end-users or other stakeholders into energy management activities a clear communication of energy utilization and market prices is necessary. This can be achieved by emerging smart home technical devices or apps for energy visualization but also be driven by legislating institutions. One example is the German smart metering legislation which ensures the (gradual) introduction of smart metering concepts with mandatory installations of smart meters in new buildings or during major renovation works. Smart metering not only enables the communication between end-users and energy utilities. It can also create a stronger awareness of energy consumption in society that potentially contributes to energy efficiency and carbon emission reduction.

6 Summary

Many different concepts are used to model smart grid systems and a high variety of approaches exists. Although the focus of the introduced model categories (large-scale, small-to-medium scale, and bottom-up demand side models) can overlap, differences exist regarding the dimensions time horizon, scope, and model perspective. Large-scale energy system models are rather applied to long-term analysis of challenges in terms of RES integration and flexibility provision. From a system perspective, the role of the energy demand side becomes more relevant in energy systems with higher shares of weather-dependent renewables. Besides the digitalization, allowing for an automated flexibilization of the energy demand, one of the main drivers of this development is sector coupling. Thus, the new additional power demands and their impacts, both on the yearly demand and on the hourly (or quarter-hourly) demand profile, have to be considered. The increase in number of actors with new technologies and the role of decentralization on the supply and demand side, emerge the importance of modeling approaches on small-to-medium scale. Since the interaction of prosumers and the respective optimization of individual or regionally aggregated energy supply and demand is located on a decentral level, a narrower geographical scope as well as time horizon is more suitable for these models. While the demand side hasn't played that crucial role in former years in energy system modeling approaches, today, a detailed bottom-up consideration is often required to address the impact of new demand technologies and its control. The corresponding disaggregation of demand (and supply) data in these demand side models enables the assessment of potentials of flexible demand to integrate higher shares of RES in a short- to long-term perspective.

Further key trends of smart grid modeling can be recognized due to the increasing role of policy and society in participating in future energy systems. To analyze the interplay of the different players, besides common modeling approaches, also new modeling approaches such as bi-level programming, and agent-based simulation will allow for more applications and thus contribute to the different research activities. Thereby, transparency and traceability are getting further in importance. In consequence, open source approaches will become standard. Additionally, the weather-dependent character of renewables necessitates on the one side a high timely resolution (as already mentioned above), but on the other side it also requires a significant improvement of forecasts. In consequence, weather forecasting will increasingly gain in importance. It has to be mentioned that complexity in modeling smart grid systems will continue to increase, also because hard- and software will further develop and allow for solving larger and even more complex models. Finally, when modeling smart grids, the trade-off between model complexity and additional insights with regard to the research question at hand must be considered.

Review Question

- What are the three main drivers which necessitate smart grid modeling?
- List five elements that are often considered in model-based representations of smart grids.
- Name and explain the three dimensions of energy system model (ESM). How are they commonly modeled in smart energy system models (ESMs)?
- Classify smart energy supply models (ESMs) by means of a visualization, considering the three dimensions of ESMs, and distinguish them from "traditional" ESMs.
- Why is the importance of modeling demand increasing for smart grids?
- Why and how is modeling of energy systems affected by uncertainties resulting from weather forecasts?
- Explain the idea of bi-level programming in the context of smart grid modeling.

References

Abrell J, Kunz F (2015) Integrating intermittent renewable wind generation - a stochastic multi-market electricity model for the European electricity market. Netw Spat Econ 15(1):117–147

Adeoye O, Spataru C (2019) Modelling and forecasting hourly electricity demand in West African countries. Appl Energy

Adika CO, Wang L (2013) Autonomous appliance scheduling for household energy management. IEEE Trans Smart Grid 5(2):673–682

Arabali A, Ghofrani M, Etezadi-Amoli M, Fadali MS, Baghzouz Y (2012) Genetic-algorithm-based optimization approach for energy management. IEEE Trans Power Deliv 28(1):162–170

Bazan P, Pruckner M, Steber D, German R (2015) Hierarchical simulation of the German energy system and houses with PV and storage systems. In: Gottwalt S, König L, Schmeck H (eds) Energy informatics. Springer International Publishing, Cham, pp 12–23

Boßmann T, Elsland R, Lickert F, Wietschel M (2013) The german load curve in 2050: structural changes through energy efficiency measures and their impacts on the electricity supply side. In: ECEEE summer study proceedings

Boßmann T, Staffell I (2015) The shape of future electricity demand: exploring load curves in 2050s Germany and Britain. Energy

Bottaccioli L, Estebsari A, Patti E, Pons E, Acquaviva A (2017) A novel integrated real-time simulation platform for assessing photovoltaic penetration impacts in smart grids. In: Energy procedia, vol 111, pp 780–789. Elsevier Ltd

Brijs T, Stiphout AV, Siddiqui S, Belmans R (2017) Evaluating the role of electricity storage by considering short-term operation in long-term planning. Sustain Energy Grids Netw 10:104–117

Bylling HC (2018) Bilevel optimization with applications in energy

Capasso A, Lamedica R, Prudenzi A, Grattieri W (1994) A bottom-up approach to residential load modeling. IEEE Trans Power Syst

Capros P, et al (2016) EU reference scenario 2016 : energy, transport and GHG emissions : trends to 2050. Technical report

Cebulla F, Fichter T (2017) Merit order or unit-commitment?: How does thermal power plant modeling affect storage demand in energy system models? Renew Energy 105:117–132

Child M, Kemfert C, Bogdanov D, Breyer C (2019) Flexible electricity generation, grid exchange and storage for the transition to a 100% renewable energy system in Europe. Renew Energy 139:80–101

Chow SK, Lee EW, Li DH (2012) Short-term prediction of photovoltaic energy generation by intelligent approach. Energy Build 55:660–667

Conejo AJ, Morales JM, Baringo L (2010) Real-time demand response model. IEEE Trans Smart Grid 1(3):236–242

Connolly D, Lund H, Mathiesen BV (2016) Smart Energy Europe?: The technical and economic impact of one potential 100 % renewable energy scenario for the European union. Renew Sustain Energy Rev 60:1634–1653

Da Silva PG, Ilić D, Karnouskos S (2013) The impact of smart grid prosumer grouping on forecasting accuracy and its benefits for local electricity market trading. IEEE Trans Smart Grid 5(1):402–410

Das UK, Tey KS, Seyedmahmoudian M, Mekhilef S, Idris MYI, Van Deventer W, Horan B, Stojcevski A (2018) Forecasting of photovoltaic power generation and model optimization: a review. Renew Sustain Energy Rev 81:912–928

Elsland R, Boßmann T, Hartel R, Gnann T, Genoese M, Wietschel M (2013) Analysis of structural changes of the load profiles of the German residential sector due to decentralized electricity generation and e-mobility. Smart Innovation, Systems and Technologies

Fernández-Blanco Carramolino R, Careri F, Kavvadias K, Hidalgo-Gonzalez I, Zucker A, Peteves E (2017) Systematic mapping of power system models. Expert survey. Technical report

Fleiter T, Worrell E, Eichhammer W (2011) Barriers to energy efficiency in industrial bottom-up energy demand models - a review

Gillingham K, Newell RG, Pizer WA (2008) Modeling endogenous technological change for climate policy analysis. Energy Econ

Gils HC, Scholz Y, Pregger T, Tena DLD, Heide D (2017) Integrated modelling of variable renewable energy-based power supply in Europe. Energy 123:173–188

Gottwalt S, Gärttner J, Schmeck H, Weinhardt C (2016) Modeling and valuation of residential demand flexibility for renewable energy integration. IEEE Trans Smart Grid 8(6):2565–2574

Goude Y, Nedellec R, Kong N (2013) Local short and middle term electricity load forecasting with semi-parametric additive models. IEEE Trans Smart Grid 5(1):440–446

Hainoun A (2009) Construction of the hourly load curves and detecting the annual peak load of future Syrian electric power demand using bottom-up approach. Int J Electr Power Energy Syst

Hakimi SM, Moghaddas-Tafreshi SM (2014) Optimal planning of a smart microgrid including demand response and intermittent renewable energy resources. IEEE Trans Smart Grid 5(6):2889–2900

Hayn M, Zander A, Fichtner W, Nickel S, Bertsch V (2018) The impact of electricity tariffs on residential demand side flexibility: results of bottom-up load profile modeling. Energy Syst

Heitsch H, Römisch W (2003) Scenario reduction algorithms in stochastic programming. Comput Optim Appl 24(2–3):187–206

Herbst A, Reiter U, Rehfeldt M, Elsland R, Fleiter T (2017) Benchmarking the EU reference scenario 2016: an alternative bottom-up analysis of long-term energy consumption in Europe. In: ECEEE summer study proceedings, pp 159–169

Herbst A, Toro F, Reitze F, Jochem E (2012) Introduction to energy systems modelling. Swiss J Econ Stat

Hidalgo Gonzalez I, Quoilin S, Zucker A (2014) Dispa-SET 2.0: Unit commitment and power dispatch model. Technical report

Hinz F (2017) Voltage stability and reactive power provision in a decentralizing energy system: dissertation

Hobbie H, Schmidt M, Möst D (2019) Windfall profits in the power sector during phase III of the EU ETS: interplay and effects of renewables and carbon prices. J Clean Prod 240(2019)

IAEA (2006) Model for Analysis of Energy Demand (MAED-2). Technical report, International Atomic Energy Agency, Vienna

Iria JP, Soares FJ, Matos MA (2018) Trading small prosumers flexibility in the energy and tertiary reserve markets. IEEE Trans Smart Grid 10(3):2371–2382

Kahrobaee S, Rajabzadeh RA, Soh L-K, Asgarpoor S (2012) A multiagent modeling and investigation of smart homes with power generation, storage, and trading features. IEEE Trans Smart Grid 4(2):659–668

Koch M, Bauknecht D, Heinemann C, Ritter D, Vogel M, Tröster E (2015) Modellgestützte Bewertung von Netzausbau im europäischen Netzverbund und Flexibilitätsoptionen im deutschen Stromsystem im Zeitraum 2020–2050. Zeitschrift für Energiewirtschaft 39(1):1–17

Kuo P-H, Huang C-J (2018) An electricity price forecasting model by hybrid structured deep neural networks. Sustainability 10(4):1280

Ladwig T (2018) Demand Side Management in Deutschland zur Systemintegration erneuerbarer Energien : Dissertation

Lee H, Eom J, Cho C, Koo Y (2019) A bottom-up model of industrial energy system with positive mathematical programming. Energy

Lorenz E, Heinemann D, Kurz C (2012) Local and regional photovoltaic power prediction for large scale grid integration: assessment of a new algorithm for snow detection. Prog Photovolt Res Appl 20(6):760–769

Lund H, Kempton W (2008) Integration of renewable energy into the transport and electricity sectors through V2G. Energy Policy 36(9):3578–3587

Luo Y, Zhu T, Wan S, Zhang S, Li K, Ev B-S, Ev F-C (2016) Optimal charging scheduling for large-scale EV (electric vehicle) deployment based on the interaction of the smart-grid and intelligent-transport systems. Energy 97:359–368

Mathiesen BV, Lund H, Connolly D, Wenzel H, Østergaard PA, Möller B, Nielsen S, Ridjan I, Karnøe P, Sperling K, Hvelplund FK (2015) Smart energy systems for coherent 100 % renewable energy and transport solutions. Appl Energy 145:139–154

Mohsenian-Rad A-H, Leon-Garcia A (2010) Optimal residential load control with price prediction in real-time electricity pricing environments. IEEE Trans Smart Grid 1(2):120–133

Morrison R (2018) Energy system modeling: public transparency, scientific reproducibility, and open development. Energy Strategy Rev 20:49–63

Möst D, Fichtner W, Grunwald A (2009) Energiesystemanalyse

Möst D, Keles D (2010) A survey of stochastic modelling approaches for liberalised electricity markets. Eur J Oper Res 207(2):543–556

Müller T, Möst D (2018) Demand response potential: available when needed? Energy Policy 115(December 2017):181–198

Nitsch J, Pregger T, Scholz Y, Naegler T, Heide D, Luca De Tena D, Trieb F, Nienhaus K, Gerhardt N, Trost T, Von Oehsen A, Schwinn R, Pape C, Hahn H, Wickert M, Sterner M, Wenzel B (2012) Long-term scenarios and strategies for the deployment of renewable energies in Germany in view of European and global developments Summary of the final report. Technical report

Oei PY, Herold J, Mendelevitch R (2014) Modeling a carbon capture, transport, and storage infrastructure for Europe. Environ Model Assess 19(6):515–531

Ottesen SØ, Tomasgard A, Fleten S-E (2016) Prosumer bidding and scheduling in electricity markets. Energy 94:828–843

Paatero JV, Lund PD (2006) A model for generating household electricity load profiles. Int J Energy Res

Panos E, Lehtilä A (2016) Energy technology systems analysis programme dispatching and unit commitment features in TIMES

Pina A, Silva C, Ferrão P (2011) Modeling hourly electricity dynamics for policy making in long-term scenarios. Energy Policy

Pipattanasomporn M, Feroze H, Rahman S (2009) Multi-agent systems in a distributed smart grid: design and implementation. In: 2009 IEEE/PES power systems conference and exposition, pp 1–8. IEEE

Poulin A, Dostie M, Fournier M, Sansregret S (2008) Load duration curve: a tool for technico-economic analysis of energy solutions. Energy Build

Ramchurn SD, Vytelingum P, Rogers A, Jennings N (2011) Agent-based control for decentralised demand side management in the smart grid. In: The 10th international conference on autonomous agents and multiagent systems-volume 1, pp 5–12. International Foundation for Autonomous Agents and Multiagent Systems

Richardson I, Thomson M, Infield D, Clifford C (2010) Domestic electricity use: a high-resolution energy demand model. Energy Build

Ringkjøb HK, Haugan PM, Solbrekke IM (2018) A review of modelling tools for energy and electricity systems with large shares of variable renewables

Riva F, Gardumi F, Tognollo A, Colombo E (2019) Soft-linking energy demand and optimisation models for local long-term electricity planning: an application to rural India. Energy

Rodriguez F, Fleetwood A, Galarza A, Fontán L (2018) Predicting solar energy generation through artificial neural networks using weather forecasts for microgrid control. Renew Energy 126:855–864

Rosen J (2007) The future role of renewable energy sources in European electricity supply A model-based analysis for the EU-15. Doctoral thesis, Universität Fridericiana zu Karlsruhe

Schill WP, Zerrahn A, Kunz F (2017) Prosumage of solar electricity: pros, cons, and the system perspective

Schönheit D, Möst D (2019) The effect of offshore wind capacity expansion on uncertainties in Germany's day-ahead wind energy forecasts. Energies 12(13):2534

Schuller A, Flath CM, Gottwalt S (2015) Quantifying load flexibility of electric vehicles for renewable energy integration. Appl Energy 151:335–344

Shang C, Wei P (2018) Enhanced support vector regression based forecast engine to predict solar power output. Renew Energy 127:269–283

Sobri S, Koohi-Kamali S, Rahim NA (2018) Solar photovoltaic generation forecasting methods: a review. Energy Convers Manag 156:459–497

Stadler M, Krause W, Sonnenschein M, Vogel U (2009) Modelling and evaluation of control schemes for enhancing load shift of electricity demand for cooling devices. Environ Model Softw 24(2):285–295

Suganthi L, Samuel AA (2012) Energy models for demand forecasting - a review

Tao L, Gao Y (2020) Real-time pricing for smart grid with distributed energy and storage: a non-cooperative game method considering spatially and temporally coupled constraints. Int J Electr Power Energy Syst 115

TEP (2020) Building stock model (BSM)

van de Ven PM, Hegde N, Massoulié L, Salonidis T (2013) Optimal control of end-user energy storage. IEEE Trans Smart Grid 4(2):789–797

Voulis N, Warnier M, Brazier FMT (2017) Impact of service sector loads on renewable resource integration. Appl Energy

Vytelingum P, Voice TD, Ramchurn SD, Rogers A, Jennings NR (2010) Agent-based micro-storage management for the smart grid

Wang F, Li K, Zhou L, Ren H, Contreras J, Shafie-Khah M, Catalão JPS (2019) Daily pattern prediction based classification modeling approach for day-ahead electricity price forecasting. Int J Electr Power Energy Syst 105:529–540

Worrell E, Ramesohl S, Boyd G (2004) Advances in energy forecasting models based on engineering economics. Annu Rev Environ Resour

Yu W, An D, Griffith D, Yang Q, Xu G (2015) Towards statistical modeling and machine learning based energy usage forecasting in smart grid. ACM SIGAPP Appl Comput Rev 15(1):6–16

Yu Z, Jia L, Murphy-Hoye MC, Pratt A, Tong L (2013) Modeling and stochastic control for home energy management. IEEE Trans Smart Grid 4(4):2244–2255

Zepter JM, Lüth A, del Granado PC, Egging R (2019) Prosumer integration in wholesale electricity markets: synergies of peer-to-peer trade and residential storage. Energy Build 184:163–176

Zerrahn A, Schill W-P (2015) A greenfield model to evaluate long-run power storage requirements for high shares of renewables

Zhang J-L, Zhang Y-J, Li D-Z, Tan Z-F, Ji J-F (2019) Forecasting day-ahead electricity prices using a new integrated model. Int J Electr Power Energy Syst 105:541–548

Ziel F, Weron R (2018) Day-ahead electricity price forecasting with high-dimensional structures: univariate vs. multivariate modeling frameworks. Energy Econ 70:396–420

Zöphel C, Schreiber S, Möst D, Herbst A, Klingler A-L, Manz P, Heitel S, Wyrwa A, Raczyński M, Fermi F, Reiter U, Jakob M (2019) Deliverable D4.3 report on cost optimal energy technology portfolios for system flexibility in the sectors heat, electricity and mobility. Technical report

Smart Grid Analytics

Christoph M. Flath and Nikolai Stein

Learning Objectives

- Grasp the basic properties of big data and how they relate to smart grids
- Understand the different facets of business analytics in the context of the power industry
- Obtain a working knowledge of machine learning tasks and algorithms
- Apply analytics knowledge to smart grid use cases

The rapid digitization of the electricity sector has lead to a massive increase of data availability. Smart grids produce large volumes of data, thanks to IoT devices like smart meters. This necessitates appropriate Data Analytics capabilities to tap into the envisioned opportunities (Zhang et al. 2018). Against the backdrop of ubiquitous computing, companies are building up capabilities for data analysis as well as automatic decision-making. Such analytics platforms can analyze data to generate findings that lead to several benefits, like cost reduction and operational efficiency. Varaiya et al. (2011) note that sensors and smart meters will provide system operators with more detailed information on the power system state. This may even lead to several other improvements in grid optimization and customer engagement. With respect to the consumption data, electricity providers can use the revealed information patterns to improve their business processes.

C. M. Flath (✉) · N. Stein
Chair of Information Systems and Business Analytics, Julius-Maximilians-Universität Würzburg, Sanderring 2, 97070 Würzburg, Germany
e-mail: christoph.flath@uni-wuerzburg.de

N. Stein
e-mail: nikolai.stein@uni-wuerzburg.de

C. Dinther et al. (eds.), *Smart Grid Economics and Management*,
Lecture Notes in Energy 51, https://doi.org/10.1007/978-3-030-84286-4_7

The idea of "business analytics" generating business value from data has emerged in the last decade. Depending on the specific focus and impact of the application, one can distinguish descriptive, predictive, and prescriptive analytics (Lustig et al. 2010). Following this classification, *descriptive analytics* encompasses business intelligence tools and processes that use data to understand and analyze past business performance. It focuses on reporting and visualization of historical data and is thus decoupled from future decisions. In contrast to this backward-oriented application, *predictive analytics* aims at uncovering explanatory structures in the data to draw inferences about future instances. The most encompassing application is *prescriptive analytics* which integrates insights from descriptive and predictive analytics to determine appropriate actions or decisions by means of optimization.

Given the increasing adaption of smart grid technologies it is hardly surprising that researchers are underlining the central importance of data mining and analytics techniques for the transformation of the electricity system (Keshav and Rosenberg 2011; Ramos and Liu 2011).

1 Big Data in Smart Grids

Solving problems and answering questions through data analysis has emerged as standard practice across many industries undergoing the digitization trend. The energy industry is no exception to this general trend (Gust et al. 2017). There are numerous rapidly evolving technologies for analyzing data and building models. This section serves to establish fundamental concepts and relate them to the energy industry.

1.1 Properties of Big Data

While the attribute "big" is a very subjective characterization of a data source. However, there is a general consensus that there are four central attributes that characterize big data analytics. Frequently, these are referred to as the four Vs: volume, variety, velocity, and veracity (Demchenko et al. 2014; Dong and Srivastava 2013; Kaisler et al. 2013; Sagiroglu and Sinanc 2013).

The main characteristic that renders a dataset "big" is its size, i.e., *Volume*. It is important to stress that size cannot be referred to in absolute quantities since the information and the capability to process it is growing exponentially every year. Ultimately, the definition of big data depends on whether the data can be ingested, processed, and timely examined to meet an organization's requirements. In smart grids IoT devices are placed in different areas like the substations and consumer devices. These devices collectively produce petabytes of data across various timescales and aggregation levels. In turn, it is not possible to make sense of the data without smart grid analytics capabilities.

Variety is one the most interesting developments in technology as more and more information is digitized. Traditional data types (*structured data*) include things on a bank statement like date, amount, and time. These are things that fit neatly in a relational database. To put it generally, structured data is well-defined, featuring a set of rules which facilitate queries and reasoning on the data. (e.g., a database schema). This stands in stark contrast to *unstructured data*, which includes any data sources that cannot be described by such a meta model. Typical examples include text, audio, and visual information (images or video). This lack of structure of these data assets prevents straightforward analysis and in turn necessitates major pre-processing efforts. Handling unstructured data assets is a fundamental challenge in big data analytics. One of the goals of big data analytics is to use technology to take this unstructured data and make sense of it. In smart grids we encounter both structured and unstructured data. Structured data will typically include various forms of time series data (e.g., consumption, generation, prices) while unstructured data may emerge in the form of textual data (e.g., maintenance reports or customer requests), graph data (e.g., grid topologies or building plans), images (e.g., satellite images). Consequently, comprehensive smart grid analytics capabilities necessitate various forms of big data analysis (Zhang et al. 2018).

With so much data available, ensuring its relevance and quality differentiates between success and failure in big data analytics. This is reflected in the *Veracity* dimension which reflects the reliability of data assets. Can decision-makers trust the data to guide their course of action? In complex systems, various discrepancies may emerge—typical examples include duplicates, outliers, inconsistencies, or volatility. Smart grid nodes like smart meter are numerous and decentral and will in individually be exposed to service interruptions leading to unreliable aggregate data. Consequently, smart grid consumption data is likely to suffer from inconsistencies and missing data which necessitates appropriate detection and mitigation strategies (Chen et al. 2017).

Velocity is the frequency of incoming data that needs to be processed. Considering the sheer number of messages, file uploads, or website interactions taking place on a particular social network site offers a good appreciation of the importance of velocity. In the context of energy markets data velocity behaves in lock-step with the velocity of the underlying market transactions. Day-ahead trading means daily updates while real-time markets will necessitate continuous data exchange. The same reasoning applies to metering infrastructure where decision-makers are striking a balance between the investment cost burden, coordination capabilities, and privacy challenges of different smart metering time resolutions (Eibl and Engel 2014; Feuerriegel et al. 2016; Flath 2013).

1.2 Data Sources in the Smart Grid

Prior to dwelling on analytics applications, it is of central importance to assess the available data assets. This includes the systems involved (data sources) and the properties of the data streams. A natural approach to structure these data assets is to align them with the electricity value chain (Zhang et al. 2018): Generation → Transmission and Distribution → Consumers. Thereby we highlight both data sources as well as potential data analytics use cases.

Generation Data

In the narrowest sense power generation data can be distinguished into output time series and capacity state information (location, technology, ramping characteristics) with varying aggregation level (ranging from generation unit to transmission system). Such information is key for any system-related analysis concerning, e.g., market prices, emissions, or availability. Notably, generation data pertains to all dimensions of the energy trilemma cost-sustainability-reliability (Heffron et al. 2015). In particular, the increased adoption of RES has resulted in previously unknown levels of short-term output fluctuations and volatility. Consequently, short-term power generation forecasts have become much more important. These play a key role in operational decisions of conventional power plants, inform trading decisions, as well as triggering demand response preparations. Given the strong correlation between RES output and the corresponding weather phenomena the relevant metrics can also be considered important generation data (e.g., solar irradiance, wind speed, water levels). More recently, researchers have even started to consider building orientation, shading conditions or roof surfaces to assess generation potentials (Fitriaty and Shen 2018).

Grid Data

Building upon a collaboration with a Swiss utility company Gust et al. (2016) highlight the different sources and forms of electricity grid data. On the one hand there is topographic and spatial data which is typically accessed through a geographic Information System (GIS). On the other hand there is investment and equipment information recorded in corporate asset databases embedded enterprise resource planning (ERP) systems. GIS data typically includes component locations of grid assets as well as information on the service area (e.g., infrastructure such as buildings and roads). Asset databases provide detailed information on asset types, their electrical specifications and thereby serve as a master data provider. Furthermore, investment cost and depreciation schedules are provided by ERP systems.

Consumption Data

With annual meter readings for many consumers electricity consumption data has traditionally been the blind spot of electricity systems. To compensate the absence of live data, planners and operators have relied on statistical models and static load profiles. These are primarily based on contracting data and the customers' historic consumption quantities (per billing period). Yet, in the last decade the idea of smart metering has increasingly challenged this myopic aggregation of (annual) consumption values. This new addition to the energy policy toolbox can have a marked effect on household energy management in turn creating change agents for the challenge of reducing (peak) demand and the integration of renewable generation (Darby 2010).

1.3 *From Descriptive to Prescriptive Analytics*

Analytics is defined by INFORMS as "the scientific process of transforming data into insight for making better decisions" (INFORMS 2020). With the rise of big data, analytics has received more attention. The most common categorization in the field is the distinction between descriptive, predictive, and prescriptive analytics (Lustig et al. 2010):

- **Descriptive analytics** relies on large amounts of historical data to identify patterns and to illustrate past behavior (den Hertog and Postek 2016). Here, a data set S^N containing information on quantities of interest (e.g., demand) $Y^N = \{\mathbf{y}^1, \ldots, \mathbf{y}^N\}$ as well as auxiliary data on associated covariates (features) $X^N = \{\mathbf{x}^1, \ldots, \mathbf{x}^N\}$ where $\mathbf{x}^t$ is concurrently observed with $\mathbf{y}^t$ are aggregated and analyzed. These observations are used for reporting, dashboards, and data visualizations. These applications are summarized in the umbrella term Business Intelligence (Watson 2014; Delen and Ram 2018). Business Intelligence systems automatically retrieve and process data from various data sources to support decision-making.
- **Predictive analytics** aims at informing decision-makers by predicting future scenarios (Corea 2016). To this end, we seek to use statistical learning to find a function $h^*(\mathbf{x}) \in H$ that minimizes a given (typically symmetric) loss function $\mathcal{L}$. In predictive analytics this loss is broadly speaking a measure of the predictive power (the smaller the loss the better the predictiveness). The set of permissible functions H is governed by the selected learning algorithm:

$$h^*(\mathbf{x}) = \underset{h(\mathbf{x}) \in H}{\arg\min}\, \mathcal{L}\left(h(\mathbf{x}), \mathbf{y}\right). \tag{1}$$

  Decision-makers use predictive analytics to generate forecasts on future scenarios $s = h^*(\mathbf{x})$ that allow them to take informed decisions even in uncertain situations.
- **Prescriptive analytics** aims at providing decision support or decision automation. In many applications the way to derive optimal policies from forecasts is not straightforward. Here, prescriptive analytics has the potential to build on operations management techniques such as mathematical optimization and decision rules to determine optimized policies. To achieve this, one no longer chooses the loss function ℓ such that it tracks predictive power but rather it emerges from the business problem (e.g., lost profit or additional cost) underlying a decision $\mathbf{q}$. One way to establish such a loss function is by comparing the outcome of a prescribed action in a given scenario to the theoretically optimal profit $\Pi^*(\mathbf{q}^*, \mathbf{y})$:

$$\ell\left(\mathbf{q}, \mathbf{y}\right) = \Pi^*(\mathbf{q}^*, \mathbf{y}) - \Pi(\mathbf{q}, \mathbf{y}). \tag{2}$$

  One can than use the forecasts obtained from predictive analytics as input parameters for optimization models to find optimal policies sequentially. Alternatively, ML algorithms can be used to find a function $q^*(\mathbf{x})$ that maps directly from data to decisions by minimizing the expected lost profit (Eq. 2):

$$q^*(\mathbf{x}) = \arg\min_{q(\mathbf{x}) \in H} \frac{1}{N} \sum_{(\mathbf{x},\mathbf{y}) \in S} \ell\left(q(\mathbf{x}), \mathbf{y}\right). \tag{3}$$

In summary, predictive analytics applications deal with forecasting future scenarios whereas prescriptive analytics prescribes a decision in anticipation of the future (Bertsimas and Kallus 2020).

2 Machine Learning Fundamentals

Many algorithms and methods for big data analytics (predictive as well as prescriptive) originate from the field of machine learning. Such algorithms do not rely on hard-coded rules but instead *learn* the underlying patterns and regularities in a data-driven fashion from provided training data. By and large this "learning activity" corresponds to an optimization of the underlying loss function.

In this context one typically distinguishes two distinct learning tasks: unsupervised learning and supervised learning (Fig. 1). As the name suggests, unsupervised learning is machine learning tasks in which humans do not need to supervise the model. With respect to the training data unsupervised learning deals with unlabeled data, i.e., there are no annotations of a possible target variable. Instead, the model works independently to discover information and patterns that were previously undetected. In contrast, supervised learning is the task of inferring a function from labeled training data. Here, each training example is a pair consisting of an input object (in most cases a vector) and a output value. Problems with a continuous output space are further categorized under the term regression problems while classification describes problems with a discrete output space. In between these archetypes there are some further learning paradigms such as semi-supervised learning or reinforcement learning.[1] However, in the context of this book we focus on the two basic cases.

2.1 Unsupervised Learning

Unsupervised learning summarizes machine learning algorithms that aim at revealing hidden structures in unlabeled data (Hastie et al. 2009). The main goal of these algorithms is either dimensionality reduction or cluster analysis.

Dimensionality Reduction

Working in high-dimensional spaces can be undesirable for different reasons—in particular processing time and storage space requirements, multi-collinearity issues,

[1] Semi-supervised learning uses a small amount of labeled data bolstering a larger set of unlabeled data. In a reinforcement learning setting the system is not trained on labels but instead on rewards resulting from interaction with the environment.

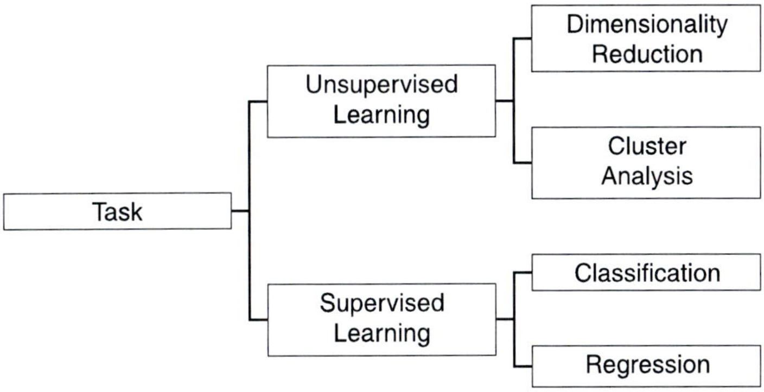

Fig. 1 Taxonomy of machine learning tasks

difficulty of visualizing and communicating data beyond two or three dimensions, and avoiding the curse of dimensionality. For these reasons dimensionality reduction is a standard step in data pre-processing. It is a transformation of data from a high-dimensional space into a lower dimensional space where the reduced representation retains the key properties of the original data. The principle component analysis (PCA) is a popular and well-studied method to transform high-dimensional datasets into low-dimensional datasets. PCA converts a set of observations of possibly correlated variables into a set of values of linearly uncorrelated variables. To this end, it finds the n principal axes in the original m-dimensional space where the variance between the points is the highest. By selecting the axes that explain most of the variance, the number of variables is reduced from m to n. Thereby, the bulk of information is preserved as the new variables are combinations of the old variables (Hotelling 1933). However, PCA reaches its limitations if the relationships between the variables are non-linear. This shortcoming is tackled by more recent approaches (Maaten and Hinton 2008).

Clustering and Segmentation

Cluster Analysis is used to group and identify data-inherent structures. These algorithms assign data points to limited number of clusters to group similar data points. The basic premise of the clustering optimization task can be summarized as follows: data points assigned to the same cluster should be as similar as possible (within-cluster homogeneity) while data points from different clusters should not be similar (across-cluster heterogeneity). From this requirement it becomes obvious that one needs two components to formulate the underlying optimization problem:

- A suitable distance function that allows to measure the similarity of two data points (greater similarity corresponds to a shorter distance).
- A loss function that aggregates the distance measurements to evaluate the quality of the cluster configuration as a whole.

Any suitable metric can be used as a distance function. Typical candidates include Euclidean distance or cosine similarity. A straightforward measure for within-cluster homogeneity h^w is given by the following expression:

$$h^w = \sum_{n=1}^{N}\sum_{k=1}^{K} r_{nk} \, ||\mathbf{x}_n - \mu_k|| \, . \tag{4}$$

Here, N is the number of data points, K is the number of clusters, $||\cdot||$ is the chosen metric, and μ_k is the central point of cluster k (centroid). The binary variable r_{nk} indicates whether or not n is in cluster k. Between cluster heterogeneity h^b is easily measured by adjusting the h^w formula which measures the total distance of the cluster members to the centroid.

$$h^b = \sum_{i=1}^{N}\sum_{k=1}^{K} (1 - r_{nk}) \, ||\mathbf{x}_i - \mu_k|| \tag{5}$$

.

Overall clustering quality CQ can then quantified by the ratio of the two aggregates:

$$CQ = \frac{h^w}{h^b} . \tag{6}$$

Note that a lower CQ score indicates a better clustering result in the sense of a low h^w value (high within-cluster homogeneity) and a high h^b value (high between-cluster heterogeneity).

There are many different algorithms available in modern software packages to execute cluster analysis task. Depending on the clustering process, we can distinguish between hierarchical algorithms (Ward Jr 1963; Murtagh and Contreras 2012) and partitioning algorithms (MacQueen et al. 1967; Kaufman and Rousseeuw 2009). In the case of hierarchical clustering the various clusters form a nested hierarchy (individual clusters are subsets of their super-ordinate clusters). Conversely, partitioning algorithms yield non-overlapping clusters.

2.2 *Supervised Learning*

Supervised learning is the machine learning task of inferring a function from labeled training data (Mohri et al. 2012). In the typical supervised learning setting we have to predict unknown future variables (e.g., energy load, renewable generation) based on past observations to make informed decisions. While experienced decision-makers may have a good intuition on some factors (e.g., weather, holidays) influencing the unknown variable, it is hard for humans to quantify the impact of all possible factors. Supervised learning approaches provide a methodological way of quantifying such influences. Depending on the underlying quantity we distinguish between *regression* and *classification*:

Table 1 Confusion matrix for the evaluation of binary classifier

Realization		Prediction: true	Prediction: false
	true	true pos. TP	false neg. FN
	false	false pos. FP	true neg. TN

Regression

These problems aim to learn a mathematical relation $f : \mathbb{R}^n \mapsto \mathbb{R}$ explaining a continuous dependent output variable y in terms of n independent variables $\mathbf{x}$. The function f allows us to obtain an estimate for the dependent variable from its associated independent variables $\hat{y}_i = f(\mathbf{x}_i)$. In the supervised learning context the loss function in the training data is obtained from comparing estimates $\hat{y}_i$ with the true values y_i. Typical candidates include the mean squared error (MSE) $\sum_i \left(\hat{y}_i - y_i\right)^2$ or the mean absolute deviation (MAD) $\sum_i \left|\hat{y}_i - y_i\right|$. Typical regression use cases include forecasting of loads, RES generation, and power exchange prices.

Classification

Many quantities of interest (e.g., fault occurrence in smart grid systems, identification of household loads) cannot be represented by continuous output variables. In such classification problems, we want to predict an observation's class y based on the feature variables $\mathbf{x}$. Here, we try to find a function $f : \mathbb{R}^n \mapsto \{1, \ldots, k\}$ where n describes the number of features and k the number of different classes. The evaluation of classification performance is best illustrated using a confusion matrix for binary classification (Table 1). This matrix compares the predictions of a classifier against the known realizations. Depending on the prediction and the realization there are four possible outcomes—true positive (TP, the classifier correctly predicts a true), true negative (TN, the classifier correctly predicts a false), false positive (FP, the classifier incorrectly predicts a true), and false negative (FN, the classifier incorrectly predicts a false).[2]

Based on the confusion matrix we can derive some standard quality metrics of the classifier:

[2] For non-binary classification the confusion matrix generalizes in a straightforward manner to a $n \cdot n$ square where n is the number of cases.

- Accuracy—the percentage of correct predictions $ACC = \frac{TP+TN}{TP+FP+TN+FN}$
- True positive rate (sensitivity)—the proportion of positive cases that are correctly identified $TPR = \frac{TP}{TP+FN}$
- True negative rate (specificity)—the proportion of negatives cases that are correctly identified $TNR = \frac{TN}{TN+FP}$
- Balanced Accuracy—the average of sensitivity and specificity which provides a better overall assessment of classifier quality in settings with unbalanced data $\frac{TPR+TNR}{2}$
- Matthews Correlation Coefficient—As suggested this coefficient is indeed a correlation coefficient between the observed and predicted binary classifications. It is generally regarded as a balanced measure which can be used even if the classes are of very different sizes. $\text{MCC} = \frac{TP\times TN-FP\times FN}{\sqrt{(TP+FP)(TP+FN)(TN+FP)(TN+FN)}}$.

Given the discreteness of the confusion matrix these quality metrics are typically non-differentiable which makes them unsuitable for serving as a machine learning loss. Instead, one typically leverages probability space and uses appropriate transformations like logistic or hinge loss.

Classification tasks in the power sector include asset reliability monitoring and prediction, customer churn analysis, or identification of appliances from meter readings.

Algorithms

We can distinguish between parametric and non-parametric machine learning algorithms to perform classification and regression tasks. Parametric algorithms are efficient to train and easy to interpret as they make strong assumptions on the form of the function. Therefore, this class of models is also referred to as white-box models. For regression tasks, multiple linear regression is the most prominent white-box algorithms. This method was established by Galton (1886) over 200 years ago and is probably the best studied form of statistical learning (Hastie et al. 2009). It rests on the assumption of a linear relationship between a set of input variables and a single output variable. The linearity assumption is simultaneously the major strength and weakness of this method. On the one hand, it renders the model very simple to understand and efficient to learn as well as robust to outliers. On the other hand, it constrains the predictive power of the model as many statistical relations are non-linear.

In contrast, non-parametric models do not make strong assumptions about the relationship between independent and dependent variables. Therefore, they are free to learn any functional form from the training data. While this flexibility oftentimes allows non-parametric algorithms to achieve a higher performance, more historical training data is required to avoid over-fitting. Non-parametric machine learning algorithms are also referred to as black-box models as the potentially higher performance comes at the cost of interpretability. The SVM is a prominent example of non-parametric algorithms. Here, each observation is viewed as a vector of all input variables. During the model training the hyper-plane that best separates the different output variables depending on the input vectors is determined. To incorporate non-linear relationships, the dimensionality of the input vector is augmented using the kernel-trick if no hyper-plane separating all different output variables exists. A major benefit of support vector machines is their good generalization ability and therefore the low susceptibility to over-fitting even for small training data sets (Smola and Schölkopf 2004). In recent years, deep neural networks have shown remarkable performance across different tasks and therefore attracted attention in research and practice. This class of models is inspired by the distributed communication and information processing in biological systems. These models consist of several layers of artificial neurons. Each neuron is connected with many other neurons and processes incoming information and propagates the results to other neurons. In the third class of black-box model we introduce gradient boosting machines. This model class combines many weak learners to a strong predictive model in a sequential fashion. This way it is able to alleviate some of the shortcomings of the weak models while increasing the predictive power. On the downside, gradient boosting is prone to over-fitting (Hastie et al. 2009).

3 Application Examples

3.1 Load and Generation Forecasting

Hong and Fan (2016) from the very beginning of the electric power industry the problem of load forecasting has been central to the business. This is of course due to the fundamental balancing requirements of the power system. Hence, power output from conventional power plants was planned in accordance with forecast load. With increasing levels of generation from fluctuating RES an additional challenge in the energy mix is the stochasticity of the supply side. This results in the novel challenge of net load forecasting, that is the difference between intermittent renewable generation and fluctuating load (Wang et al. 2017). This quantity ultimately determines whether measures aimed at decreasing net load are needed (additional generation capacities have to be ramped, shifting of flexible loads) or alternatively measures to increase net load are to be performed (charging of storage facilities, expanding flexible consumption, ramping down generators).

Time Series Analysis Forecasting

The classic forecasting approach in power systems management is time series analysis (Amjady 2001; Espinoza et al. 2005). Time series analysis leverages past observations of a variable to predict its future values. A time series in this context is a sequence of values over time (daily, monthly,...) of a quantitative variable, $\langle y_t, y_{t-1}, y_{t-2}, ...\rangle$. Future realizations of the quantity of interest are assumed to be a function of the past realizations and some white noise random error term. Consequently future values can be predicted is $\hat{y}_t = f(y_{t-1}, y_{t-2}, y_{t-3} \ldots)$. The functional mapping should capture typical underlying patterns such as trends (growth or decline) and seasonalities. The classic Holt-Winters triple exponential smoothing approach provides a powerful yet tractable framework for such problems (Chatfield 1978). The underlying logic of this procedure is decomposition of the time series into three distinct exponentially smoothed units, the level term E_t, the trend term T_t and the seasonality term S_t:

$$\hat{y}_{t+n} = (E_t + nT_t)\, S_{t+n-p}, \tag{7}$$

where

$$E_t = \alpha \left(\frac{Y_t}{S_{t-p}} \right) + (1 - \alpha)(E_{t-1} + T_{t-1}) \tag{8}$$

$$T_t = \beta (E_t - E_{t-1}) + (1 - \beta)\, T_{t-1} \tag{9}$$

$$S_t = \gamma \left(\frac{Y_t}{E_t} \right) + (1 - \gamma)\, S_{t-p} \tag{10}$$

Using this system of variables it is straightforward to choose the three parameters such that the training forecasting error is minimized. This approach is particularly suited for short-term forecasting applications (Amjady 2001). Forecasting using time series models has a long tradition in power systems management and is a robust and interpretable approach. A key disadvantage is the lack of non-time series features as these models rely solely on the information embedded in the historic time series but cannot readily incorporate other data sources. Furthermore, the relationships are restricted to the time series model and hence fairly restrictive (in particular with respect to functional relationships or the number of seasonal terms).

Machine Learning for Forecasting

The limitations of the time series analysis lend themselves to advanced analytics approaches. From a ML perspective forecasting tasks boil down to straightforward regression tasks where the target variable y_t corresponds to the load, generation, or net load at time t. Independent of the concrete use case typical features $\mathbf{x_t}$ include weather information, lagged y values (time series features), pricing data, or customer information. These are readily incorporated into standard algorithms. A particularly illustrative one is decision trees where the feature space is partitioned into groups through a sequence of (typically) binary decisions. An example in the load forecasting context is provided in Fig. 2. This example highlights how decision trees naturally incorporate rich relationships across Various features: the main driver of load in this case is the weather with high temperatures leading to a high load forecast (2,800MW, e.g., due to air conditioning) while in the case of lower temperatures other criteria like type of day or season of the year are embedded in the tree as well leading to differentiated forecasts (2,550MW for summer weekend, 2,300MW for other weekends, and 2,600MW for a working day).

Constructing a decision tree rests on a very simple procedure: Divide data using a beforehand chosen splitting rule into disjoint subsets. Repeat this algorithm recur-

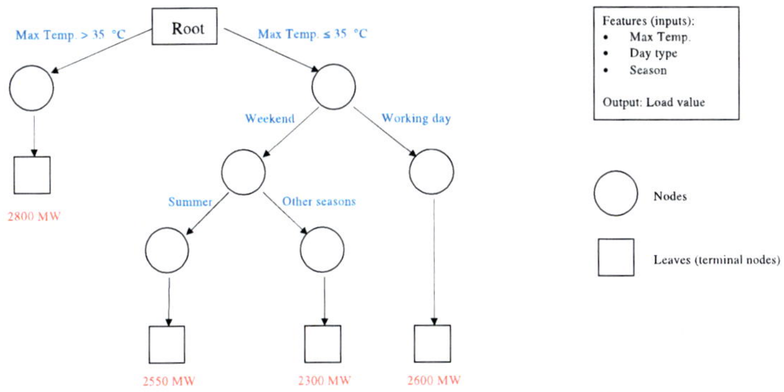

Fig. 2 Regression tree for load forecasting (Lahouar and Ben Hadj Slama 2015)

sively for each subset, stop the recursion when the leaves of the given subset are sufficiently pure.[3] Despite offering high interpretability decision trees suffer from a pronounced tendency to overfit to training data leading to subpar out of sample performance. To avoid this regularization techniques like limiting the tree depth or the number of leaves can be employed.

Going beyond a single tree ensemble methods which combine multiple decision trees helps to greatly improve the predictive performance. The two main ensembling approaches are *bootstrap aggregation* (bagging) and *boosting*:

- A bagging ensemble trains many parallel decision trees (resulting in so-called random forests Breiman 2001). The different trees are trained on datasets which are created from the original data through sampling with replacement. This parallelism increases the reliability of the predictions.
- A boosting ensemble trains many decision trees in sequence. Each tree is trained to minimize the loss taking into account the combined and weighted decision of all preceding trees. This sequence of weak learners is capable of greatly improving the overall result of the decision tree forecast (Chen and Guestrin 2016).

Recently, these machine learning approaches have seen ready adoption in the power systems literature (Persson et al. 2017; Su and Liu 2018; Guo et al. 2021).

Future Challenges in Load Forecasting

Historically, most models focused on point load forecasting. This was because there was sufficient slack in the system (due to controllable generation) which facilitated reliable matching between supply and demand. Recently, though network management has become more challenging given recent developments in the sector. Among others, probabilistic load forecasting is better suited to serve utilities companies in this challenging task (Hong and Fan 2016). This is another advantage of machine learning techniques which ultimately are inherently probabilistic. Given sufficient

[3] Purity can, for example, be measured using the misclassification error in a leaf. Alternative metrics include the Gini coefficient as well as the entropy.

training data there have also been successful applications of deep neural networks, in particular recurrent neural networks (Wang et al. 2019).

3.2 *Clustering Load Profiles*

The widespread deployment of smart metering has made household energy consumption data more available. This has paved the way for analyzing load characteristics using this advanced metering data. Such power usage data can be leveraged by system operators or energy retailers to better understand customer behavior. This information may help tailor different DR programs for different households in order to build a more robust grid design system, offer more effective energy reduction recommendations, and improve smart pricing models (Salah et al. 2017). These new sources of business value are crucial as smart meters are more costly than traditional metering equipment as they require large investments in the accompanying ICT infrastructure. Consequently, utilities companies are intensively looking for ways to recapture these investments. Faruqui et al. (2010) remark that utilities can profit from improved grid operations, lower costs for obtaining meter readings, or faster identification of outages and interruptions. Smart metering data can facilitate a customer segmentation based on dynamic load patterns instead of mere load totals. Flath et al. (2012) present a practice-oriented introduction to smart meter cluster analysis which we want to quickly outline this user case of unsupervised learning here.

Time Series Preparation

To be able to cluster a collection of smart meter time series (load profiles) we need to bring these data into a usable form. Traditional customer segments often focus on the total consumption as this data has always been readily available from billing systems. Clearly, this corresponds to a very coarse definition of similarity. A smart meter data series with a typical resolution of 15 min resolution provides $4 \cdot 24 \cdot 365 = 35,040$ data points per household. The central question is the underlying clustering objective, i.e., what measures similarity between two load profiles. To compare two load profiles side-by-side it is first necessary to determine the comparison scope and then obtain corresponding representative load profiles. Flath et al. (2012) put forward daily (differentiated by weekday and weekend days) and weekly profiles as natural candidates. In either case the quarter-hourly raw data are aggregated at the chosen level by means of averaging. This results in load profiles with 96 entries for daily and 672 entries for weekly profiles—for weekly profiles it may hence be advisable for weekly profiles to be based on coarser time resolution. To obtain scale-free comparability across load profiles one can normalize the values by division with the highest individual element.

The representative load profiles offer compact representation of the customer consumption records. Pairwise comparison of profiles requires furthermore a suitable metric. Given the time series character of the load data we can simply use a default metric such as the Euclidean metric.

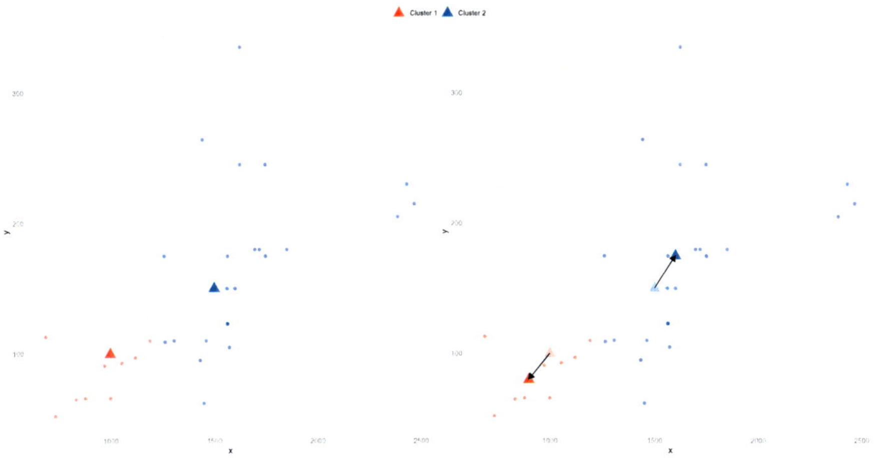

Fig. 3 Illustration of k-means clustering approach

Clustering Procedure

Equipped with representative load profiles and a suitable metric one can proceed to the actual clustering process. A very simple and commonly used algorithm for cluster analysis is the k-means method. This algorithm requires a pre-defined number of clusters, k. It then proceeds by picking k representative load profiles as cluster centers and subsequently assigns all remaining load profiles to the closest cluster. Subsequently, the cluster centers are recalculated and the cluster assignments are updated. This procedure is repeated until there are no more assignment changes which then yields the final clustering. See Fig. 3 for an illustration of the first two steps of the algorithm.

Given the random initialization and the greedy improvement procedure k-means can sometimes end up in sub-optimal clusterings. This issue can be mitigated by trying out multiple starting configurations. The other inherent issue of k-means is the necessity of specifying the number of clusters ex-ante. This can be overcome with additional ex-post metrics of cluster quality to compare across different k-values. Note that for this comparison the number of clusters has to be included as an additional distance penalty as the plain clustering quality is strictly increasing in the number of clusters. Note that oftentimes determining an "optimal" number of clusters is a somewhat theoretical question as practical considerations (cost of marketing additional products, marketing strategy) may anyway impose a fixed value for k.

Advanced Clustering Techniques

The above example is deliberately used the k-means approach as it is the most common and approachable technique for cluster analysis. However, it is somewhat limited with respect to unpacking underlying behavioral aspects of the consumers. To this end, Albert and Rajagopal (2013) propose using a hidden Markov model to understand the inter-temporal consumption dynamics to appropriately segment the user population. They infer occupancy states from consumption time series data and characterize occupancy by magnitude, duration, and variability. Besides its non-probabilistic nature, the k-means approach also struggles to incorporate other data sources than the consumption time series as pointed out by Wang et al. (2018).

Putting Customer Segments to Use

More granular load profiles provide ample benefit potentials for the various stakeholders in smart grids (Beckel et al. 2014). For example, energy companies can better manage their assets and customer relationships. In particular, the cluster assignment can be a powerful feature in the context of load forecasting as described above. Furthermore, marketing activities can be better targeted with respect to both customers targeting as well as product design. Customers themselves may receive better recommendations and comparisons which may improve their consumption behavior. Finally, regulators may be able to turn richer insights into customer behavior into more effective legislation and rules for the energy market.

3.3 Non-intrusive Load Monitoring

A central theme of smart grids is the empowerment of more local units, i.e., pushing planning and control paradigms from the TSO to the DSO level. This coincides with greater visibility on individual consumers. Yet, these customers remain are still represented as a load profile which of course is a poor representation of the underlying energy end use. This is where load disaggregation techniques such as Non-intrusive load monitoring (NILM) come into play. NILM processes meter data in order to determine exactly what appliances have used the power and how much power each appliance has used during that time period (Hamid et al. 2017). This challenging class of smart grid analytics problems features elements of supervised multi-class classification (identification known appliance traces in an aggregate load profile) as well as unsupervised discovery of previously unseen loads. Weiss et al. (2012) highlight the particular challenge of reliably detecting inductive or capacitive loads. Unsurprisingly, research has so far been driven by the technology and less so by the resulting business opportunities (Welikala et al. 2017). It is also the main application area of advanced AI techniques as the sequence-based and super positional nature of aggregated load profiles is distinctly non-structured. Consequently, ANN approaches such as sequence (Zhang et al. 2018) or transfer learning (D'Incecco et al. 2019) have recently been used to tackle this problem.

If the NILM problem can be reliably solved the opportunities for future smart grid management would be significant. On the customer end there would be a great improvement of information transparency with regards to individual energy consumption. This may help identify out-dated appliances or bad consumption habits. Companies seeking to establish comprehensive DR programs could reliably target customers with certain appliance types or usage patterns. Similarly, customer response and hence the effectiveness of RTP or ToU rate offerings could greatly benefit from knowing what customers actually do with the procured energy.

4 Summary

As discussed throughout this book the smart grid is a transformation of the power system both from the technological as well as the organizational perspective. Smart grid analytics assumes a similar perspective by highlighting the necessity of first turning data sources into information assets and subsequently leveraging those to generate new business value.

The increasing availability of smart grid data has lead to significant research activity in the smart grid analytics domain. However, so far a lot of research has focused on descriptive and predictive modeling tasks. Going forward one could expect a closer integration of these predictive ML models with smart grid optimization problems (e.g., Chen et al. 2013; Gärttner et al. 2018; Gust et al. 2021) to establish a truly data-driven prescriptive decision-making paradigm. An early pioneer in this area has been Rudin et al. (2011) who cooperated with a New York City utility company to schedule maintenance crews based on reliability predictions (predictive maintenance). The adoption of the weighted sample average approximation (Bertsimas and Kallus 2020) may provide a promising framework for tackling the prescriptive challenge in smart grid analytics more broadly.

NILM research has presented an avenue for deep learning applications. Another area where these deep neural networks are of particular relevance in smart grids is image recognition which can be used to extract information on grid assets (type, size or condition) or customer demand from aerial or satellite imagery (Yan et al. 2007; Malof et al. 2015; Gazzea et al. 2021).

Review Question

- Name the four Vs of big data and elaborate on each of them in the context of smart grid analytics.
- Explain the difference between prescriptive and predictive analytics in your own words.
- Random forests are often considered to be unpractical for large data sets due to their high memory usage. Explain this critique by means of a stylized example (1GB training data, 10 trees).
- Why is the CQ metric not useful for choosing the number of clusters?
- Discuss how electricity pricing and rate design can benefit from smart grid analytics.
- What type of machine learning task do the examples for image recognition correspond to?

References

Albert A, Rajagopal R (2013) Smart meter driven segmentation: what your consumption says about you. IEEE Trans Power Syst 28(4):4019–4030

Amjady N (2001) Short-term hourly load forecasting using time-series modeling with peak load estimation capability. IEEE Trans Power Syst 16(3):498–505

Beckel C, Sadamori L, Staake T, Santini S (2014) Revealing household characteristics from smart meter data. Energy 78:397–410

Bertsimas D, Kallus N (2020) From predictive to prescriptive analytics. Manag Sci 66(3):1025–1044

Breiman L (2001) Random forests. Mach Learn 45(1):5–32

Chatfield C (1978) The holt-winters forecasting procedure. J R Stat Soc Ser C (Appl Stat) 27(3):264–279

Chen T, Guestrin C (2016) Xgboost: a scalable tree boosting system. In: Proceedings of the 22nd ACM SIGKDD international conference on knowledge discovery and data mining, KDD '16, pp 785–794

Chen W, Wang X, Petersen J, Tyagi R, Black J (2013) Optimal scheduling of demand response events for electric utilities. IEEE Trans Smart Grid 4(4):2309–2319

Chen W, Zhou K, Yang S, Wu C (2017) Data quality of electricity consumption data in a smart grid environment. Renew Sustain Energy Rev 75:98–105

Corea F (2016) What data science means to the business. In: Big data analytics: a management perspective, vol 21. Springer International Publishing, Cham, pp 5–17

Darby S (2010) Smart metering: what potential for householder engagement? Build Res Inf 38(5):442–457

Delen D, Ram S (2018) Research challenges and opportunities in business analytics. J Bus Anal 1(1):2–12

Demchenko Y, De Laat C, Membrey P (2014) Defining architecture components of the big data ecosystem. In: 2014 international conference on collaboration technologies and systems (CTS). IEEE, pp 104–112

den Hertog D, Postek K (2016) Bridging the gap between predictive and prescriptive analytics-new optimization methodology needed. Tilburg Univ, Tilburg, The Netherlands

D'Incecco M, Squartini S, Zhong M (2019) Transfer learning for non-intrusive load monitoring. IEEE Trans Smart Grid 11(2):1419–1429

Dong XL, Srivastava D (2013) Big data integration. In: 2013 IEEE 29th international conference on data engineering (ICDE). IEEE, pp 1245–1248

Eibl G, Engel D (2014) Influence of data granularity on smart meter privacy. IEEE Trans Smart Grid 6(2):930–939

Espinoza M, Joye C, Belmans R, De Moor B (2005) Short-term load forecasting, profile identification, and customer segmentation: a methodology based on periodic time series. IEEE Trans Power Syst 20(3):1622–1630

Faruqui A, Harris D, Hledik R (2010) Unlocking the 53 euro billion savings from smart meters in the EU: how increasing the adoption of dynamic tariffs could make or break the eu's smart grid investment. Energy Policy 38(10):6222–6231

Feuerriegel S, Bodenbenner P, Neumann D (2016) Value and granularity of ict and smart meter data in demand response systems. Energy Econ 54:1–10

Fitriaty P, Shen Z (2018) Predicting energy generation from residential building attached photovoltaic cells in a tropical area using 3d modeling analysis. J Clean Prod 195:1422–1436

Flath CM (2013) An optimization approach for the design of time-of-use rates. In: IECON 2013-39th annual conference of the ieee industrial electronics society. IEEE, pp 4727–4732

Flath CM, Nicolay D, Conte T, van Dinther C, Filipova-Neumann L (2012) Clusteranalyse von Smart-Meter-Daten. Wirtschaftsinformatik 54(1):1–10

Galton F (1886) Regression towards mediocrity in hereditary stature. J Anthropol Inst G B Irel 15:246–263

Gärttner J, Flath CM, Weinhardt C (2018) Portfolio and contract design for demand response resources. Eur J Oper Res 266(1):340–353
Gazzea M, Pacevicius M, Dammann DO, Sapronova A, Lunde TM, Arghandeh R(2021) Automated power lines vegetation monitoring using high-resolution satellite imagery. IEEE Trans Power Deliv
Guo W, Che L, Shahidehpour M, Wan X (2021) Machine-learning based methods in short-term load forecasting. Electr J 34(1):106884
Gust G, Brandt T, Mashayekh S, Heleno M, DeForest N, Stadler M, Neumann D (2021) Strategies for microgrid operation under real-world conditions. Eur J Oper Res 292(1):339–352
Gust G, Flath C, Brandt T, Ströhle P, Neumann D (2016) Bringing analytics into practice: evidence from the power sector. In: Proceedings of the 37th international conference on information systems
Gust G, Neumann D, Flath CM, Brandt T, Ströhle P (2017) How a traditional company seeded new analytics capabilities. MIS Q Exec 16(3):215–230
Hamid O, Barbarosou M, Papageorgas P, Prekas K, Salame C (2017) Automatic recognition of electric loads analyzing the characteristic parameters of the consumed electric power through a non-intrusive monitoring methodology. Energy Procedia 119:742–751
Hastie T, Tibshirani R, Friedman J (2009) The elements of statistical learning, 2nd edn
Heffron RJ, McCauley D, Sovacool BK (2015) Resolving society's energy trilemma through the energy justice metric. Energy Policy 87:168–176
Hong T, Fan S (2016) Probabilistic electric load forecasting: a tutorial review. Int J Forecast 32(3):914–938
Hotelling H (1933) Analysis of a complex of statistical variables into principal components. J Educ Psychol 24(6):417
INFORMS (2020) Best definition of analytics
Kaisler S, Armour F, Espinosa JA, Money W (2013) Big data: issues and challenges moving forward. In: 2013 46th Hawaii international conference on system sciences. IEEE, pp 995–1004
Kaufman L, Rousseeuw PJ (2009) Finding groups in data: an introduction to cluster analysis, vol 344. Wiley
Keshav S, Rosenberg C (2011) How internet concepts and technologies can help green and smarten the electrical grid. ACM SIGCOMM Comput Commun Rev 41(1):109–114
Lahouar A, Ben Hadj Slama J (2015) Day-ahead load forecast using random forest and expert input selection. Energy Convers Manag 103:1040–1051
Lustig I, Dietrich B, Johnson C, Dziekan C (2010) The analytics journey. Anal Mag, pp 11–13
Maaten Lvd, Hinton G (2008) Visualizing data using t-sne. J Mach Learn Res 9:2579–2605
MacQueen J, et al (1967) Some methods for classification and analysis of multivariate observations. In: Proceedings of the fifth Berkeley symposium on mathematical statistics and probability, Number 14. Oakland, CA, USA, , pp 281–297
Malof JM, Hou R, Collins LM, Bradbury K, Newell R (2015). Automatic solar photovoltaic panel detection in satellite imagery. In: 2015 international conference on renewable energy research and applications (ICRERA). IEEE, pp 1428–1431
Mohri M, Rostamizadeh A, Talwalkar A (2012) Foundations of machine learning. MIT press
Murtagh F, Contreras P (2012) Algorithms for hierarchical clustering: an overview. Wiley Interdiscip Rev Data Min Knowl Discov 2(1):86–97
Persson C, Bacher P, Shiga T, Madsen H (2017) Multi-site solar power forecasting using gradient boosted regression trees. Solar Energy 150:423–436
Ramos C, Liu C (2011) Ai in power systems and energy markets. IEEE Intell Syst 26(2):5–8
Rudin C, Waltz D, Anderson RN, Boulanger A, Salleb-Aouissi A, Chow M, Dutta H, Gross PN, Huang B, Ierome S et al (2011) Machine learning for the New York city power grid. IEEE Trans Pattern Anal Mach Intell 34(2):328–345
Sagiroglu S, Sinanc D (2013) Big data: a review. In: 2013 international conference on collaboration technologies and systems (CTS). IEEE, pp 42–47

Salah F, Flath CM, Schuller A, Will C, Weinhardt C (2017) Morphological analysis of energy services: paving the way to quality differentiation in the power sector. Energy Policy 106:614–624

Smola AJ, Schölkopf B (2004) A tutorial on support vector regression. Stat Comput 14(3):199–222

Su H-Y, Liu T-Y (2018) Enhanced-online-random-forest model for static voltage stability assessment using wide area measurements. IEEE Trans Power Syst 33(6):6696–6704

Varaiya P, Wu F, Bialek J (2011) Smart operation of smart grid: risk-limiting dispatch. Proc IEEE 99(1):40–57

Wang K, Qi X, Liu H (2019) Photovoltaic power forecasting based lstm-convolutional network. Energy 189:116225

Wang Y, Chen Q, Hong T, Kang C (2018) Review of smart meter data analytics: applications, methodologies, and challenges. IEEE Trans Smart Grid 10(3):3125–3148

Wang Y, Zhang N, Chen Q, Kirschen DS, Li P, Xia Q (2017) Data-driven probabilistic net load forecasting with high penetration of behind-the-meter pv. IEEE Trans Power Syst 33(3):3255–3264

Ward JH Jr (1963) Hierarchical grouping to optimize an objective function. J Am Stat Assoc 58(301):236–244

Watson HJ (2014) Tutorial: big data analytics: concepts, technologies, and applications. Commun Assoc Inf Syst 34(1):65

Weiss M, Helfenstein A, Mattern F, Staake T (2012) Leveraging smart meter data to recognize home appliances. In: 2012 IEEE international conference on pervasive computing and communications. IEEE, pp 190–197

Welikala S, Dinesh C, Ekanayake MPB, Godaliyadda RI, Ekanayake J (2017) Incorporating appliance usage patterns for non-intrusive load monitoring and load forecasting. IEEE Trans Smart Grid 10(1):448–461

Yan G, Li C, Zhou G, Zhang W, Li X (2007) Automatic extraction of power lines from aerial images. IEEE Geosci Remote Sens Lett 4(3):387–391

Zhang C, Zhong M, Wang Z, Goddard N, Sutton C (2018) Sequence-to-point learning with neural networks for non-intrusive load monitoring. In: Proceedings of the AAAI conference on artificial intelligence, vol 32

Zhang Y, Huang T, Bompard EF (2018) Big data analytics in smart grids: a review. Energy Inform 1(1):8

Business Model Design

Markus Lau, Clemens van Dinther, and Orestis Terzidis

Learning Objectives

- Understand basic terms and concepts of business models,
- Understand the opportunities arising from a new market environment,
- Identify cross-sector opportunities,
- Study current trends in the start-up sector,
- Understand and apply the different dimensions of a business model,
- Learn and apply modeling concepts and business model design with its different design phases,
- Know and apply modeling tools.

1 New Business Models and Innovation in the Smart Grid

Until a few years ago, the utility companies in the electricity sector followed a rather simple business model: electricity was generated by centrally located power plants and then supplied to as many customers as possible via a network of poles and

C. van Dinther (✉)
ESB Business School, Reutlingen University, Alteburgstr. 150, 72762 Reutlingen, Germany
e-mail: clemens.van_dinther@reutlingen-university.de

M. Lau · O. Terzidis
Institute for Entrepreneurship, Technology Management and Innovation (EnTechnon), Karlsruhe Institute of Technology (KIT), Fritz-Erler-Straße 1-3, 76133 Karlsruhe, Germany
e-mail: markus.lau@kit.edu

O. Terzidis
e-mail: orestis.terzidis@kit.edu

C. Dinther et al. (eds.), *Smart Grid Economics and Management*,
Lecture Notes in Energy 51, https://doi.org/10.1007/978-3-030-84286-4_8

wires. The key value proposition was to deliver electricity as safely, reliably, and cost-effectively as possible and to keep outages or interruptions to a minimum.

Triggered by new technologies and numerous external influences such as new political decisions, customer relations and market structure are changing more and more. In the coming years, power utilities, political decision-makers, authorities, and other stakeholders will have to reinvent themselves to meet emerging customer needs, as well as the issues of sustainability, decarbonization, electrification of transport, the introduction of distributed energy resources, and smart grids.

We will look at some of these new trends and developments leading to this reinvention and the resulting challenges and opportunities in the field of Smart Grids.

1.1 Some Basic Terms

The changing market environment creates opportunities for both, incumbents and new entrants. As the market environment changes, established companies are looking for new ways to earn money. At the same time, start-ups, which are often more agile and faster, are entering the market. Both players need to develop new business models. A business model (BM) is generally understood to be the description of an abstract business idea with its value-adding activities and processes for achieving a competitive advantage. We will discuss the concept of a BM in more detail in Sect. 2. The BM concept is closely linked to the concept of strategy. In the following, we will distinguish these different terms from each other. The overall goal of a *strategy* is to create a unique and valuable position, involving a different set of activities than its rivals, to gain superior performance in the markets within which it operates (Barney and Arikan 2001; Porter 2004). Barney and Arikan (2001) imply that this superior performance may be understood as the combination of persistent economic rent generation and sustainable competitive advantage. Further, there are different strategy hierarchy levels distinguishing between corporate strategy addressing the organization as a whole, business unit strategy inherent to the respective subordinate business unit, and functional strategies, supporting their superordinate business units through the development of functional competencies (Hax and Majluf 1991).

The BM also aims toward competitive advantage through an exceptional value proposition for customers and value creation partners (Wirtz 2010; Clauss 2017; Teece 2010). Furthermore, the BM is a multi-level concept also covering the company level, the business unit level, as well as the level of specific products and services. In principle, both concepts have similarities and thus pursue a similar goal. However, there are different streams in management literature of how these two concepts are distinct or interlinked with each other.

Seddon and Lewis (2003), therefore, also speak of two identical concepts. However, many management researchers see strategy and BMs as two different concepts due to clear differences in the consideration of competition, financing, and knowledge (Chesbrough 2002; Zott and Amit 2008). Alt and Zimmermann (2001) see the value proposition as the only overlap between strategy and BM and, therefore, speak of two independent concepts with a common intersection.

According to Magretta (2002), Teece (2010), and Zott and Amit (2008), a BM provides the value creation architecture for implementing a strategy—they see the BM as a component of the strategy. For Deelmann and Loos (2003), a BM provides the basis for the development of corporate and functional strategies, according to which the strategy is part of the BM.

In the meantime, however, a more and more homogeneous understanding of the interrelations between the two terms is becoming apparent. As already described by Al-Debei and Avison (2010), there is rather no partial but a hierarchical dependency between strategy and BM. According to Richardson (2008), the BM explains how the company's activities collaborate in implementing its strategy, thereby bridging the formulation and implementation of the strategy (Zott et al. 2011; Richardson 2008). Similarly, Shafer et al. (2005), as well as Casadesus-Masanell and Ricart (2011), describe the BM as a reflection of a company's strategy. According to Osterwalder and Pigneur (2002), the BM is an instrument for the coherent implementation of a strategy, based on which operational implementation can take place within the framework of organizational design or business process model. Wirtz et al. (2016), as well as Zott et al. (2011), understand the BM as a link between future planning (strategy) and operational implementation (business process management). Bieger and Reinhold (2011) conclude that the strategy provides the reference framework for the development and design of a BM. The scope of action of a strategy represents the starting point for the development of BMs. BMs, thus, serve to implement strategies.

A *business plan* is an entrepreneurial planning instrument (Thommen 2007). As such, it includes a written summary of an entrepreneurial project, such as setting up a business or launching a new market service. This serves as the basis for converting all information related to production, sales, and financing into financial ratios (Singler 2012). The objective is detailed liquidity planning and the determination of the return on investment in the amount and time required (Nagl 2010). For Stähler (2002), a business plan is a BM put on paper. While a BM provides concise information on conceptual aspects of a planned project, the business plan provides evidence of economic advantage (Teece 2010). The business plan, thus, serves as a basis for discussion for internal and external stakeholders (shareholders, managing directors, business partners, investors, etc.) (Singler 2012; Thommen 2007). A BM is, therefore, an input variable for the business plan. The business plan is, thus, a concept that follows the BM (Stähler 2002). A good *business idea* is a success factor for a successful company and the basis for every BM (Heinemann 2008). A business idea describes how business activity can be used to achieve superior competitive performance (Rentmeister and Klein 2003). In a BM, a business idea is analyzed, evaluated, and concretized (Malek et al. 2004). A business idea is an input variable for BM development (Bach et al. 2003). Following this understanding, the BM links the business idea with the business plan (Scheer et al. 2003).

1.2 Opportunities from a Changing Market Environment

There are several technical, economic, and legal factors in the transition of the energy system, which make changes necessary or provide opportunities for BM innovations. Some have been discussed in other chapters already. However, we summarize the most important ones. First, Renewable energy generation is characterized by *zero marginal cost*. Once the generation facilities are installed, the generation of electricity comes at zero marginal cost (apart from possible maintenance costs (e.g., for wind)). As such, in the energy-only market, renewable generation technologies are pushing conventional power plants out of the market due to their cost structure. This creates economic pressure on conventional power plant operators and makes other market structures necessary. Second, the new energy world is largely *decentralized*, as the renewable generators are installed in smaller units at the distribution grid level. In this respect, decentralized solutions are necessary, e.g., for the efficient use of locally generated energy, as well as billing models for neighborhood electricity or neighborhood solutions. Smart homes for more efficient energy use in residential houses or smart energy quarters for cross-home solutions will be required. On the commercial level (industrial as well as crafts and trade), energy management solutions will lead to higher transparency and control through automation. This requires smart grid communication and meter infrastructure and good scalability of the solutions offered. It remains to be seen to what extent standards are necessary to achieve economies of scale. On the other hand, there are highly specialized solutions that are necessary for the individual circumstances of the customer. In this respect, a high degree of customization is also necessary. These considerations alone show a wide range of possible BM approaches. Third, *Digitalization* and *connectivity* open up new technical possibilities. If the consumption side is currently still largely analog, artificially intelligent algorithms will be able to control automatically and take over other tasks in the future. Connectivity also enables the integration of devices from different manufacturers and, thus, comprehensive control and the formation of virtual units (both on the generation side as virtual power plants and on the consumption side as virtual consumers). Fourth, Renewable energies depend on the weather and are, therefore, *volatile in generation*. For reliable energy generation, solutions are needed that can cope with this volatility. In the previous chapters, it has already been shown that there are different approaches to this, e.g., storage technologies but also flexible adaptation on the consumption side. Which solution is suitable depends not least on the situational conditions. In any case, the ability to react to fluctuating generation is necessary. Forecasting methods are becoming more important to be able to react better. Flexibility on the consumption side is economically advantageous. Digital solutions will also contribute to this. Fifth, *regulatory framework conditions* are on the one hand necessary to change the energy system or result as a consequence of technological changes. Regulatory changes challenge established BMs on the one hand and provide an opportunity to establish new BMs on the other. Markets in transition hold great potential for new business ideas. Sixth, *cross-sector opportunities* are also becoming more important. For example, through BEV the transport and energy sectors are moving closer together. Intermodal transport concepts can help to balance the respective advantages and disadvantages of different modes of transport. Intelligent charging concepts are needed to make better use of the existing electricity infrastructure and at the same time to realize the greatest possible benefit for BEV owners. The heating sector and the electricity sector will also be more closely linked if the combustion of fossil fuels becomes uneconomical and is replaced by heat pumps. Here, too, there are opportunities for new BMs. These changes call for new technical and service solutions. If we look at the retail business, for example, decentralized generation results in prosumers and consumers that not only demand but also generate and feed in electricity. Consequently, retailers will sell less electricity to such prosumers. On the other hand, feeding in electricity might not realize attractive revenues. Thus, prosumers might look for better usage of self-produced electricity or they might want to sell it on a neighborhood community basis. Both call for new services offerings, e.g., local electricity usage optimization with integrating heat pumps or battery electric vehicles that are charged with locally photovoltaics (PV)-generated electricity. Consequently, new technology is needed, not only on the electricity generation side, but also with respect to better measuring electricity consumption in combination with optimization services for better or more efficient usage, toward new Walboxes for smart charging of eVehicles, or regarding better connectivity of devices and making them smart, e.g., with internet of things (IoT) technology. We also expect new tariffs as introduced in Chap. 2. Literature on BM innovation covers different areas such as solar electricity generation (Karakaya et al. 2016), energy storage (Kittner et al. 2017), electric vehicle charging (Madina et al. 2016), or energy retailers (Karami and Madlener 2018). Before we look at exemplary use cases, we will look at the current developments of start-ups in the smart grid.

1.3 Current Development of Start-ups in the Smart Grid

The transformation of the electrical value chain offers numerous challenges but also opportunities, for established but above all for newly emerging companies. In the environment of this change, great potential is, therefore, attributed to start-ups in particular, which can open up new business fields through innovative BMs and technologies.

Primarily driven by decentralization and digitization, the topic of Smart Grids is also becoming increasingly important for start-ups. An analysis of start-up activities in the field of Power supply shows that start-ups related to Smart Grids have the second-highest growth rate after Decentralized energy systems.[1] Figure 1 shows the number of start-ups between 2010 and 2018 for various sub-sectors of the electricity supply. A total of 91 start-ups related to Smart Grids were founded during the period under review.

[1] More detailed information on the data basis of this analysis can be found in Sect. 4.2.

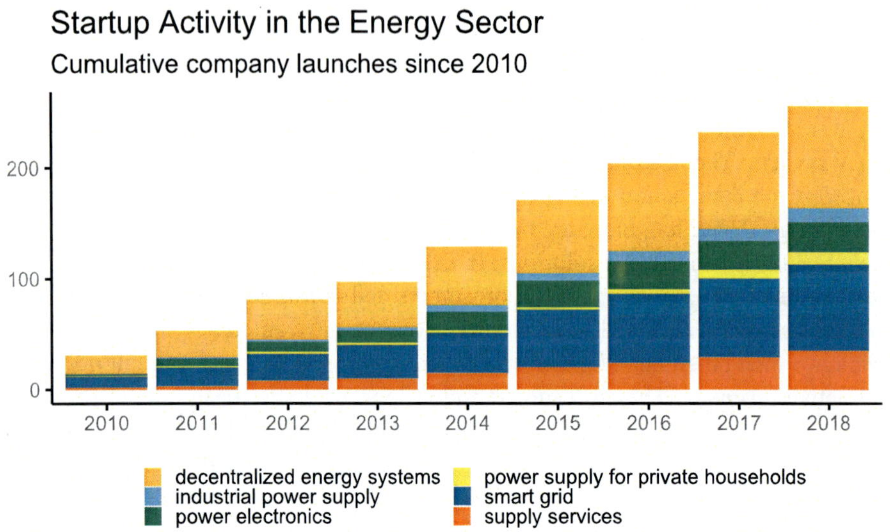

Fig. 1 Number of start-ups founded in the electricity supply sector between 2010 and 2018, Source: own calculation based on Innoloft database

A closer look at the companies' products and solutions shows a clear connection to decentralization and digitization in the Smart Grid segment. Thirty percent of Smart Grid start-ups state that they provide solutions for distribution network monitoring. 20% of the companies relate their solutions to decentralized energy systems. Just under 15% use machine learning, artificial intelligence, and cloud computing in combination with Big Data for their offerings. A wide variety of monitoring services are derived from these, such as predictive maintenance or the creation of forecasts and scenarios for generation and demand. Especially for the latter, solutions are increasingly being developed in the areas of demand-side response, management, and integration. A detailed examination of the Smart Grid start-ups also reveals the increasing networking within the value chain. Numerous start-ups offer services for companies in the power generation sector, for virtual power plants, the general pooling of generation plants, and also the pooling of consumers, for example, via peer-to-peer platforms. Solutions for the energy management of industrial companies are also integrated into the product portfolio of the start-ups. With their products and services, the start-ups contribute to the ability to offer and manage efficient and sustainable energy services in distribution networks with variable energy resources and demand points, which are already more interconnected today.

These cross-departmental solutions are also essential enablers for addressing the future energy landscape, in which distributed resources have reached a critical mass, the electrification of the transport system continues to advance, storage resources are distributed throughout the country, and flexibility is established as an economic good, to address the emerging opportunities via suitable BMs.

2 The Business Model Concept

Since the turn of the millennium, researchers and practitioners have increasingly dealt with the topic of BMs. And against the background of an increasingly globalized world and increasing customer demands, companies of all sizes are more dependent than ever on their ability to quickly adapt strategies and BMs (Heuser et al. 2007). This is why BMs gain even more popularity and becoming increasingly important as a competitive factor (Onetti et al. 2012). Every company has at least one BM, be it on the company or business unit level or be it associated with a specific product or service (Wirtz 2010). Some companies explicitly formulate their BM and communicate it openly, others keep it for themselves or act on it implicitly (Lambert 2003). But what is a BM and how is it understood in management science and practice? Let us have a look at the different levels, goals, and fields of application of the term.

2.1 The Development History

The origin of the term BM has not been clarified (Bieger and Reinhold 2011; Wirtz 2010). Although many associate the term with the rise of the new economy of the years 1998–2000 (Wirtz 2010), earlier uses of the term can be found in the literature.[2] However, it was during the years before and after the dotcom bubble in 2000, where the term BM was the central aspect of business activity for companies and their investors (Lambert 2008). Since then, the term BM found its way into the business language and management practice of the old economy as well (Wirtz 2010). Because of the widespread use of the term BM, several attempts have been made to develop a generally valid definition of the term (Al-Debei et al. 2008; MacInnes and Hwang 2003; Pateli and Giaglis 2004).[3] However, to date, no uniform definition of the term has prevailed (Wirtz et al. 2016; Foss and Saebi 2018). The heterogeneous understanding of the term, its nature, components, structure, and representation developed because different streams and different scientific disciplines have influenced and still influence the use of the term (Zott et al. 2011; Baden-Fuller and Morgan 2010; Ghaziani and Ventresca 2005; Bieger and Reinhold 2011; Wirtz 2010).

A very detailed comparison of these different streams can be found in Wirtz (2010). He traced back the different theoretical approaches for the definition of the BM concept to the three basic streams of information technology, organizational theory, and strategy theory (Wirtz 2010). Figure 2 gives an overview of the development of these approaches. The following discussion of these three streams is a summary

[2] Previous uses can be found in scientific articles such as Bellman et al. (1957), Jones (1960), and McGuire (1965); see Osterwalder et al. (2005) and Wirtz (2010).

[3] Several listings of existing definitions of the term BM can be found in the literature reviews of, e.g., Deelmann and Loos (2003), Scheer et al. (2003), Jonda (2004), Wirtz (2010), Bieger and Reinhold (2011), Zott et al. (2011), and Wirtz et al. (2016).

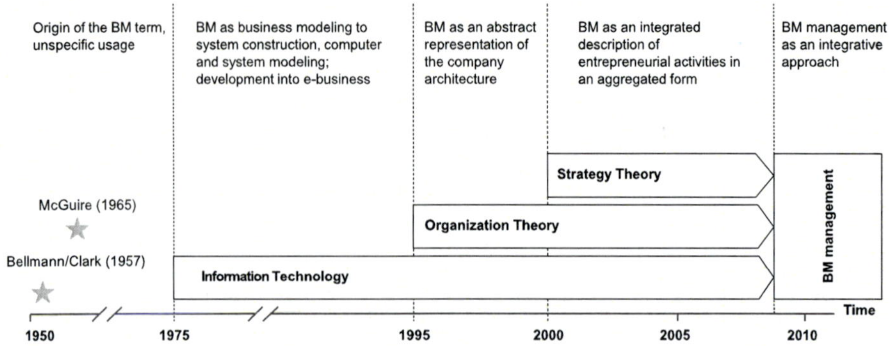

Fig. 2 Development of the BM Concept, Source: Wirtz (2010)

of the work from Wirtz (2010) and serves a better understanding of today's use of the term.

In the information technology approach, BMs originated from the research area of management information systems. The focus in this stream was on business modeling and the resulting business process model. These business processes were mapped using structured methods (e.g., UML or BPMN) to increase the efficiency and effectiveness of the information system of a company (Wirtz 2010). The BM itself formed a simplified representation of the business processes, on which the system developer constructed the information system (Wirtz 2010). Over time, however, there has been a shift in the meaning and function of BMs in the information technology approach. Instead of only describing existing processes and structures for the technical system development, the BM itself became the first step in the modeling process (Wirtz 2010). Therefore, the BM developed from a purely operative planning tool in system development to an integrated representation of the company organization as a management tool (Schoegel 2001). Since BMs were no longer restricted to the preliminary conceptual stage of systems development, this led to the development of the organization theory approach. Here, BMs developed into an independent analytical instrument that served to understand the mechanisms of companies. In this context, BMs were understood as abstract representations of the internal structure and architecture of a company. As such, the BM was seen as an important support for management decisions (Al-Debei et al. 2008).

With the functional change of the BM to a support tool for management decisions, strategy theory as another theoretical approach gained in importance (Wirtz 2010). Since 2000, more and more papers have put BMs and strategy into a close relationship, in particular, the corporate strategy (Chesbrough 2002; Kagermann and Österle 2007; Wirtz 2010). In the strategy theory approach, the BM became an aggregated description of entrepreneurial activity and provided information on which combination of production factors can be used to implement a company's business strategy (Wirtz and Kleineicken 2000). In addition to the company's internal perspective, this also led to an increase in competition-strategic components (Hamel 2002). Therefore,

BMs became a management concept for the cross-company description, analysis, and design of business activities.

2.2 *Current Understanding of the Concept*

Although these different streams exist, in literature and practice, an increasingly homogeneous understanding of the term itself and the purpose of a BM, the level of abstraction of corresponding tools, and the classification or differentiation from the strategy concept is developing (Wirtz et al. 2016; Zott et al. 2011; Burkhart et al. 2011). For example, it is now widely accepted and implicitly and explicitly recognized that the BM represents a new unit of analysis that differs from product, company, or industry (Zott et al. 2011; Burkhart et al. 2011). Further, many researchers agree that a BM is an abstract and simplified representation of the business reality or idea, like a blueprint. It describes all value-creating activities and procedures in their key business components to gain a competitive advantage (Al-Debei and Avison 2010; Doleski 2015; Wirtz et al. 2016). The concept is focused on a company, but the implications of the concept go beyond this considered company. Furthermore, many researchers acknowledge that BMs usually explain not just how customer value is created, but also how it is delivered and captured (Wirtz et al. 2016; Zott et al. 2011; Al-Debei and Avison 2010). This holistic view of a business provides a better understanding of its internal structure and functions as well as how it fits in the external environment and how it interacts with it (Al-Debei and Avison 2010). One of the most extensive attempts to connect the three streams described in the previous chapter and to develop a homogeneous understanding of the BM concept is the work of Al-Debei and Avison (2010). They try not only to unify the components of a BM but also to describe its functions and its interaction with strategy

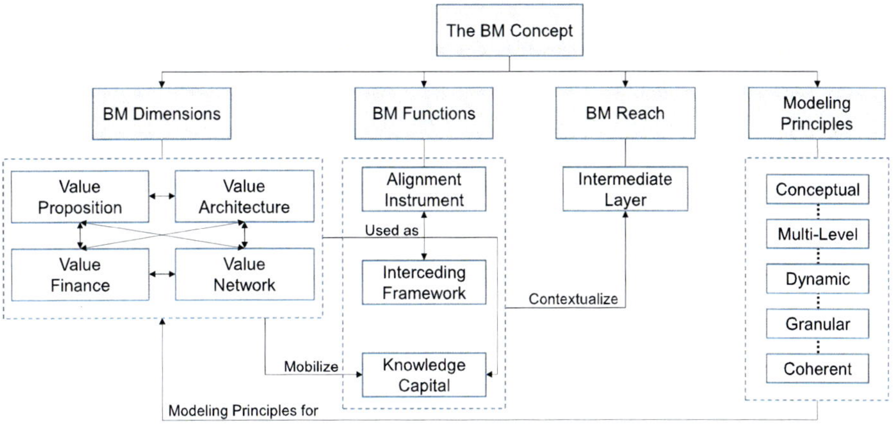

Fig. 3 A unified BM conceptual framework, Source: Al-Debei and Avison (2010)

and business processes. They also define clear modeling principles that are to serve as the basis for the development and description of a BM. This framework comprises four superordinate facets shown in Fig. 3. Rather than giving another definition of the term BM, this framework and its facets will be described in further detail to give a holistic understanding of the BM concept.

2.3 Business Model Dimensions

This facet includes four value dimensions, a value proposition for all stakeholders, a value architecture with regards to technological and organizational infrastructure, a value network to enable transactions, and the value finance dimension comprising costing, revenue, pricing, etc. These dimensions are highly interrelated and interdependent. Al-Debei and Avison (2010) see the **value proposition** as two-dimensional. On the one hand, it describes how an organization creates value with its partners and complementors for the customer. On the other hand, it can describe how the company creates value for all stakeholders involved (Al-Debei and Avison 2010; Amit and Zott 2001; Andersson et al. 2006; Magretta 2002; Osterwalder et al. 2005). In management, however, the term stakeholder is defined as all internal and external groups of persons who are directly or indirectly affected by the entrepreneurial activities at present or in the future (Freeman 2010). Following this definition, the customer groups are part of the stakeholders. The value proposition dimension of Al-Debei and Avison (2010) can, therefore, be described as more simple as how an organization creates value for all its stakeholders. This dimension includes products and services as well as the target market and customer segments. **Value architecture** focuses on the resources and competencies of an organization and their configuration with regards to value creation most effectively and efficiently (Al-Debei and Avison 2010; Barney 2010). This includes the description of the necessary internal and external activities and processes to create the value proposition. **Value finance** refers to costs, pricing methods, and revenue structures, aiming at an efficient setup to generate a beneficial financial output for an organization (Al-Debei and Avison 2010; Osterwalder et al. 2005; Shafer et al. 2005). The **value network** dimension addresses cross-company relationships within a BM (Al-Debei and Avison 2010). It includes relationships and interaction modes among stakeholders, partners, governmental agencies, or competitors (Al-Debei and Avison 2010; Bouwman 2002; Kallio et al. 2006).

2.4 Modeling Principles

Al-Debei and Avison (2010) describe the BM primarily as a **conceptual** tool of an existing or planned company. It does not cover every detail of the business logic, but focuses on the essential core elements of the business. However, it is important to understand the BM as a **coherent** concept. This means that the BM, although it does not depict all details of the business logic, nevertheless describes the company completely and comprehensively and, thus, enables a holistic view of the company. Nevertheless, this tool has **granular** properties. This means that a BM can be broken down into several controllable elements and the level of detail can, thus, be adapted to the corresponding needs of the analysis. This is particularly important if the BM has to cover a large number of business aspects. This granularity is also a prerequisite for the versatile structure of the BM. The versatile, **multi-level** feature means that a BM can be used to understand and analyze a company at different levels. A more detailed analysis of these different levels of BMs can be found in Sect. 2.7. It also means, however, that the BM can serve different purposes within a company, e.g., as an alignment instrument or knowledge capital. Another principle of the BM concept is its **dynamic** structure. This is essential since many industries themselves are exposed to very dynamic developments, to which the BM must react with the appropriate adjustments. It must, therefore, be possible to adapt dynamically and flexibly.

2.5 Business Model Reach

This facet classifies BMs as **intermediate layer**, providing an interface between business strategy and ICT-enabled business processes (see Fig. 4). This view is shared by many researchers and practitioners (Morris et al. 2005; Osterwalder et al. 2005; Al-Debei et al. 2008; Wirtz 2010; Wirtz et al. 2016). Nevertheless, it is worthwhile to differentiate between the terms strategy, BM, and business process. This is done in .

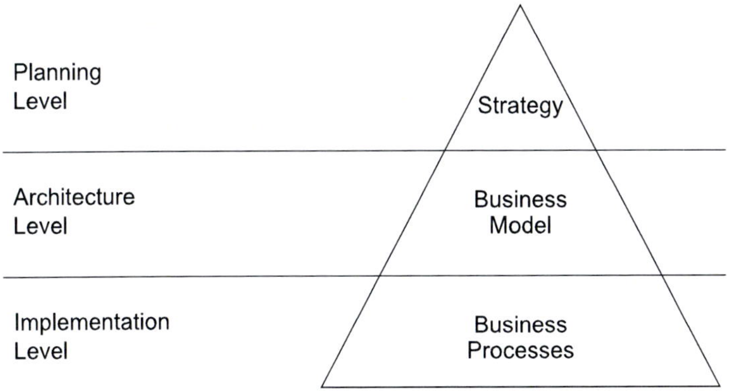

Fig. 4 Differentiation between strategy, business model, and business process (Osterwalder and Pigneur 2002; Al-Debei and Avison 2010)

2.6 Business Model Functions

This facet classifies BMs by their function. Al-Debei and Avison (2010) see the BM as a concept that serves several purposes which are not necessarily mutually exclusive. One purpose is the **alignment instrument** to harmonize strategy and business processes including information systems. This considers the greater complexity of today's industries. As already described in Sect. 2.1 on the Information Technology approach, this complexity leads to the fact that business processes can no longer be derived directly from the strategic goals of the company. The BM, therefore, serves as a tool of alignment to ensure the essential harmonization between the strategy and the processes. The BM also serves as **interceding framework** between the potentials of a technological or non-technical solution and the realization of value for the customer group. This means that technology, a product, or service alone is not enough, but can only create value in combination with a correspondingly coordinated BM. The BM is, therefore, a central component in achieving a company's strategic goals. The third purpose sees the BM as **knowledge capital** constituting an information asset to support strategic decision-making functions. Al-Debei and Avison (2010) argue that an explicitly described BM can be a critical organizational asset that enhances managers to better control the business.

2.7 Levels of Business Models

There are various abstraction levels of the concept of BMs (Schallmo and Brecht 2010; Wirtz 2010). According to Schallmo and Brecht (2010) and based on an extension of Wirtz (2010), there are two main abstraction levels of the concept of BMs comprising five sub-levels in total (see Fig. 5). These abstraction levels shall briefly be explained: Generic level: The generic level is not valid for companies due to its abstract nature and comprises the two sub-levels of abstract BM types and industry BM types:

- The abstract level is defined independently from an industry. It comprises an option space of elements and is the general principle of how to operate.
- The industry level is defined for industry and comprises an option space of elements. It is the general principle of how to act in an industry.

Specific level: The specific level is valid for companies due to its detailed nature and comprises the three sub-levels of corporate BMs, business unit models, and product and service BMs:

- The corporate level is defined for corporate business and comprises a fixed set of elements. It describes the operations of a specific company.
- The business unit level is defined for business units of corporate business and comprises a fixed set of elements. It describes the operations of a specific business unit.

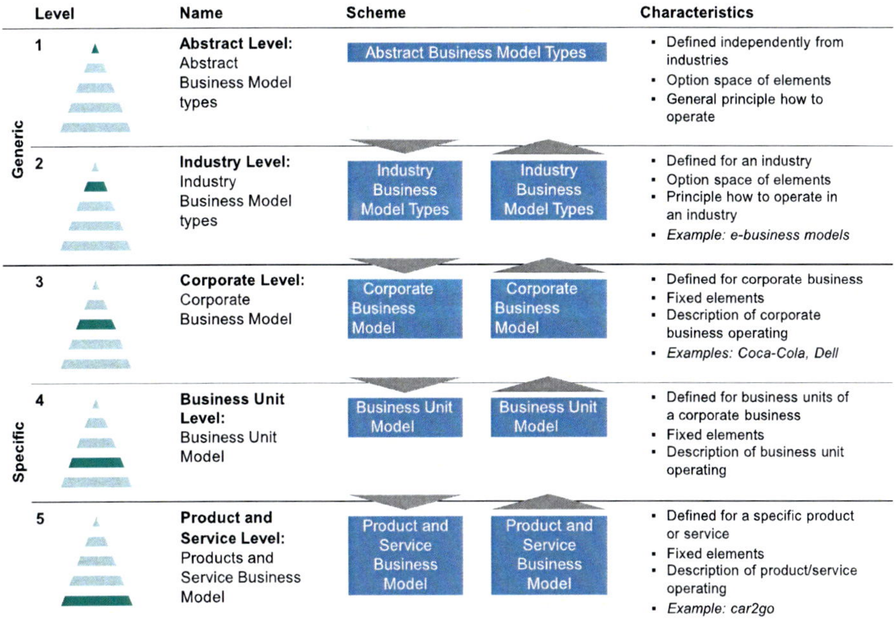

Fig. 5 BM levels according to Schallmo and Brecht (2010) and Wirtz (2010)

- The product and service level is defined for a specific product or service and comprises a fixed set of elements. It describes the operations regarding a specific product or service offering.

Depending on the context, the right level must be chosen for the presentation of the BM. This can depend on the concrete application context or, for example, on the size of the company (Wirtz 2010). If, for example, it is a matter of designing a BM for a start-up, it can be sufficient to consider the company level. With existing companies, a BM may change at the level of the business unit, but the higher level model at the company level remains unaffected. The principle of coherence is essential when considering the different levels. If changes are made at one level, it must be analyzed whether this results in changes at other levels. It is, therefore, important to understand that the levels are not mutually exclusive, but interdependent (Wirtz 2010). Thus, the company must always be viewed holistically across all levels.

2.8 *BM Goals and Areas of Application*

Following the description of the BM functions by Al-Debei and Avison (2010), BMs are used in different areas of application and, thus, pursue a variety of goals. They are used as alignment instruments to harmonize strategy and business processes including information systems, as an interceding mediating framework connecting "technological potentials and innovations with the realization of economic value and the achievement of strategic outcomes", and as knowledge capital constituting an information asset to support strategic decision-making functions. Along with these functions, BMs aim to support the analysis, design, communication, and implementation of a current or planned business activity (Bieger and Reinhold 2011; Wirtz 2010).

The overriding goal of a BM is to secure the long-term profitability and competitive advantages of the company over its competitors (Wirtz 2010; Clauss 2017; Teece 2010). In addition to this core objective, there are other objectives derived primarily from the role of the BM as an analysis and management tool. As an abstract representation of the company logic, the BM offers a comprehensive holistic understanding of the company. It, thus, comprehensively describes the business activities of a company. The abstraction level on which it is based makes it possible to simplify the management of interactions, processes, etc., and, thus, helps the management of a company to develop a better basis for decision-making. The holistic understanding combined with a high degree of abstraction provides management with the necessary information about customers, market, competition, processes, resources, competencies, and finances without going into too much detail. This representation significantly reduces the complexity of the company logic and increases the understanding of the interrelationships of the core aspects within and outside the company. The reduction of complexity and the associated simplification of decision-making are further goals of the BM (Deelmann and Loos 2004; Wirtz 2010).

The holistic understanding also aims to identify potentials better and to assess risks more precisely (Eriksson and Penker 2000). The internal and external potentials and risks have a considerable influence on a company's decisions. The identification of opportunities and risks is, therefore, an important goal of the BM for the company (Wirtz 2010).

Also, the BM serves as a communication tool to communicate the core aspects of the company to employees, investors, or other stakeholders. The BM must be understood as a uniform and concise picture of the basic mechanisms of current and planned business activities (Doganova and Eyquem-Renault 2009; Lindström 1999; Meinhardt 2002).

3 Business Model Design

BM Design is the process of designing a business activity. It serves to concretize a business idea within a strategic framework. A systematic approach can support the BM designer by guiding the creative design process along structured paths. It ensures that no important aspects of a BM are forgotten. This significantly increases the chances of success of a BM (Jonda 2004). To support the design of BMs, many practically oriented structuring approaches have been developed to make the BM concept more tangible and comparable. Also, design processes were developed to guide the entrepreneur through the creation of a BM using specific steps and methods.

For this purpose, tools were developed to support the individual process steps and to visualize the BM as a whole. These structuring approaches, the design process as well as some tools will be presented in the following chapters.

3.1 Structuring Approaches

Understanding BMs as abstract representations of business logic that describe how a company creates value for all stakeholders is the first step in designing a BM. However, this understanding is still far too imprecise for practical implementation. There are, therefore, numerous attempts to transfer the BM from theory to practice and make it more tangible for practical application. This includes, for example, attempts to describe and analyze selected BMs of companies in detail to conclude the development of new BMs. This is usually done based on case studies. Another possibility to structure the BM more strongly for practical application offers so-called taxonomies. The aim here is to describe and classify the BM based on predefined criteria. These criteria can vary from sector to sector and are usually derived from practical examples or related management concepts. Most widespread, however, is the structuring of the BM using a component-based view. This means that the BM is divided into defined components. In the literature, these components are often referred to as BM elements, dimensions, or building blocks, usually used synonymously (Osterwalder et al. 2005; Ballon 2007). There are still some major differences between the individual components, for example, the number of components included ranges from four (Value Architecture, Value Finance, Value Proposition, and Value Network) at Al-Debei and Avison (2010) to nine components (customer segments, value propositions, customer relationships, distribution channels, revenue systems, key resources, key activities, key partners, and cost structure) at Osterwalder and Pigneur (2010) or even up to 20 components at Krumeich et al. (2012). What all these approaches have in common, however, is that the components are understood throughout as interdependent elements (Burkhart et al. 2011). No matter how many components an approach ultimately contains, all these frameworks are management tools for the active development of BMs (Demil and Lecocq 2010). They provide a predefined set of components whose combination and elaboration can fully describe a BM, although in a highly simplified and aggregated way (Eurich et al. 2013; Wirtz 2010). Depending on the number of components, the level of detail and abstraction can vary according to the description. BM components consider individual core aspects of business activity, such as processes, resources, competencies, finances, and competitors (Demil and Lecocq 2010). The selection of the relevant components and the degree of information used to describe these components is largely determined by the goal of a BM (Lambert 2003). Among all frameworks, the BM Canvas by Osterwalder and Pigneur (2010) has established itself as a theoretical and practical standard, especially in the early stages of BM development. It is particularly recommended as the first starting point for entrepreneurs, as it can provide a structured overview of the essential aspects of new business without already knowing all the necessary information. This framework is described in more detail in . The component-based view of the BM is a prerequisite for another possibility to structure BMs, namely via BM patterns. BM patterns are idealized descriptions of combinations of certain components. They, thus, describe BMs with similar characteristics, similar rules for BM components, or similar behavior. A single BM can contain several of these patterns. These patterns, therefore, serve as a blueprint for the development of a BM by describing the design of particularly successful BMs in industry and revealing their similarities (Bieger and Reinhold 2011). Some patterns are limited to certain industries (e.g., e-commerce), others are also transferable to other industries. One of the best-known works on these patterns is the study of the St. Gallen BM Navigator by Gassmann et al. (2013a). In their study, they identify a total of 55 different BM patterns. According to their studies, over 90% of BM innovations are recombinations of these 55 patterns. Established patterns include Direct Selling, Long Tail, Bait and Hook, Pay-per-Use, or Freemium (Osterwalder and Pigneur 2010; Gassmann et al. 2013a). These patterns form the basis for the analysis of the most common BMs of existing energy start-ups, which is described in detail in Sect. 4.

3.2 Design Phases

The diverse approaches and advances in structuring the BM lead to the fact that literature and practice increasingly deal with processes and structures for designing BMs. Here, a distinction must first be made between two terms that are often used synonymously in this context, BM Design and BM Development. The main difference between the terms lies in the scope of the process steps. BM Design refers only to the design of the appropriate BM for a company, a business unit, or a product/service. The actual implementation of the BM is no longer part of the design process. BM Development, on the other hand, considers the entire process from the design to the actual implementation of the BM. The BM design process is, therefore, part of the BM development process. Frankenberger et al. (2013) take up this subdivision in their "4I Framework" for the BM Development Process. The framework introduces four phases for BM development. The (1) initiation phase has the goal to collect information and to conduct business context analysis as a preparation for the actual BMD. The (2) ideation phase is aiming at generating different BM ideas and concepts to create the possible solution space. In the (3) integration phase, the actual BM concepts are designed by detailing the BM idea drafts from the previous phase. In the final (4) implementation phase, the BM concept is transferred into a realization plan and executed. Within this BM development process, the BMD process is defined by phases one to three. Accordingly, the BMD process is part of the BM development process, but with the exclusion of the implementation phase (Frankenberger et al. 2013). A very similar subdivision of the BM design process can be found at Wirtz (2010). According to Wirtz (2010), the design process also starts with an initiation phase in which the initial situation is analyzed. The second process steps the idea

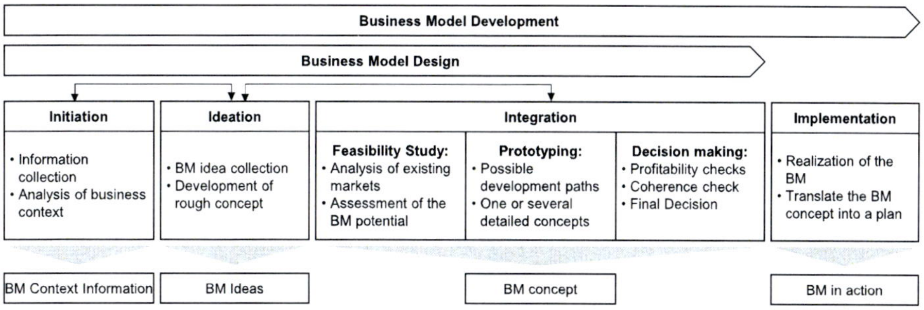

Fig. 6 BM Design process as part of the BM Development process (Frankenberger et al. 2013; Wirtz 2010)

generation which equals the ideation phase of Frankenberger et al. (2013). Ideas are collected using creativity techniques and an initial rough concept is created to assess the feasibility of the idea. Wirtz (2010) then divides the process step of the integration phase at Gassmann et al. (2013b) into three sub-steps, the feasibility analysis, the prototyping, and the decision-making. In the feasibility analysis, the existing markets are analyzed in detail to assess the potential of the BM. This includes, for example, the analysis of the customer segment, the resulting market potential, or the competition. In prototyping, various development paths of the BM are conceptualized and analyzed. From this, the individual components of the BM are worked out finely. These fine concepts serve as the basis for decision-making. For example, profitability calculations are used to identify weaknesses in detail. Besides, the components are checked for coherence and if necessary refined or harmonized. Finally, a decision on the BM is made based on an evaluation of the individual detailed concepts. This decision concludes the design process. Figure 6 combines the two approaches of Frankenberger et al. (2013) and Wirtz (2010).

Both approaches have in common that they define BM design as an iterative process. If relevant information is missing in the ideation phase, or if further analysis is needed to develop alternative solutions, a step back to the initiation phase may be important. Likewise, in the design of detailed concepts in the integration phase, an extension of the ideas from the ideation phase may occur. Only after completion of the first three phases, the implementation phase is triggered based on a BM concept (Frankenberger et al. 2013). The implementation phase is not part of the iterative BMD. Thus, once a BM is set, the activities are in place, and the resources have been developed, it is difficult to change (Zott and Amit 2007).

3.3 Business Model Design Tools

In practice, the BM design process is often supported by tools that are intended to accompany the company or entrepreneur through the process. The main task of these tools is to collect the essential information and findings and, if possible, to visualize

Table 1 List of the BM tools aggregated in categories

Category	Tool (Reference)
Generic Tool	Business Model Canvas (Osterwalder and Pigneur 2010)
	Lean Canvas (Maurya 2012)
	Business Model Navigator (Gassmann et al. 2017)
	Business Model Cube (Lindgren and Rasmussen 2013)
	Integrated Business Model (Wirtz 2010)
	RCOV Framework (Demil and Lecocq 2010)
	Innovation Readiness Levels Framework (Evans and Johnson 2013)
	Process Model of Change for Sustainability (Roome and Louche 2015)
Environment	Framework for Business Models for renewable energies (Engelken et al. 2016)
	Ecopreneurial Business Model Framework -Jolink and Niesten (2013)
	The triple layered Business Model canvas (Joyce and Paquin 2016)
	Business Model Gaming (Laurischkat and Viertelhausen 2017)
Mobility	Morphological Boxes (Kley et al. 2011)
E-Business	Dynamic Business Model Framework (Reuver et al. 2009)
	E-Business-Model-Generator (Kollmann and Hensellek 2016)

them. Thus, the complexity of the design of the BM should remain manageable. Meanwhile, there are numerous tools in the literature, some of which are generic tools for all industries and branches, others are designed for very specific industries. Table 1 shows a selection of these generic tools as well as some industry-specific tools.

As already introduced in Sect. 3.1, the BM Canvas by Osterwalder and Pigneur (2010) is one of the most widespread tools in practice and has established itself as a standard tool for BM design, especially in the field of Entrepreneurship Education. According to Osterwalder and Pigneur (2010), a BM describes the value created for selected customer segments and the value creation architecture required to create, market and distribute this value. The value represents the profitable and sustainable source of revenue of the BM. The value creation architecture is the cost-causing factor of a BM. Osterwalder and Pigneur (2010) describe a BM using nine components, the so-called "9 building blocks" (see Fig. 7).

The **value proposition** describes the benefit that a customer can derive from the company's offering. It describes which customer wishes are satisfied or which customer problems are solved and, thus, represents the reason why a customer decides in favor of the company's offer.

The **customer segments** define the individual customer groups that are to be addressed by the product. The potential customers are segmented through market analysis to be able to satisfy the different needs optimally.

The **channels** are used to determine how communication with customer groups is to take place. Communication channels, as well as distribution and sales channels, are defined.

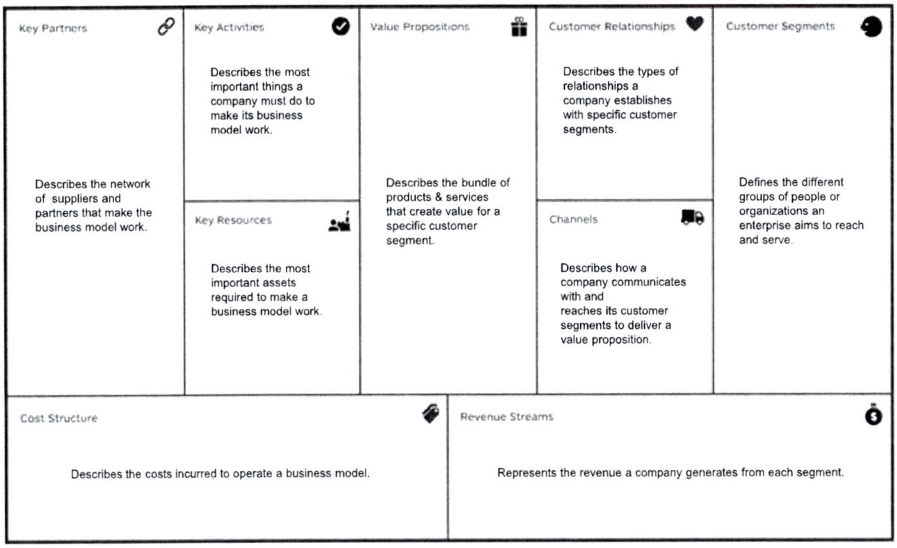

Fig. 7 The Business Model Canvas, Source: (Osterwalder and Pigneur 2010)

The **customer relationships** describe the type of relationship to be established, the degree of intensity, and the costs of establishing and maintaining it.

The **revenue structure** is used to illustrate how a company generates revenue. For this purpose, the willingness to pay of the selected customer segments is to be determined and the appropriate means of pricing is to be selected.

The **key activities** are those activities through which the formulated value proposition is delivered.

Key resources are those resources that are fundamental to the implementation of the key activities.

Key Partners characterize a network of suppliers and cooperation partners of a company, who, e.g., provide key resources.

The **cost structure** reflects the essential costs that arise from the realization of a BM. To determine this structure, the selected key activities, resources, and partnerships of the BM must be used.

Osterwalder and Pigneur (2010) group these nine building blocks into four main categories, covering the four main areas of a company: Value proposition, customer interface, infrastructure management, and financial aspects. The elements of customer segments, marketing channels, and customer relationships form the customer interface. To expand these, key resources, activities, and partners are needed that represent the infrastructure of the BM. The value proposition acts as an interface between the customer and the infrastructure. It forms the core of a BM. The financial model, consisting of the revenue sources and the cost structure, is the monetary perspective on the BM (Osterwalder and Pigneur 2010).

This generic introduction to the topic of BMs serves the understanding, use, and practical implementation of the concept as such. The following chapter examines how the concept is implemented in the energy sector. To this end, start-ups in the energy industry environment are examined concerning their BMs and these are clustered based on various BM patterns. Thus, an overview of the current developments of young companies in the energy industry will be given before a detailed analysis of selected start-ups will take place in Sect. 9.1.

4 Business Model Patterns of Energy Start-ups

The energy sector, especially in Germany, can be seen as an example of an industrial change triggered by the increased importance of sustainability. By enacting the legislative package for the energy transition in summer 2011, the Federal Government of Germany paved the way from a centrally dominated toward a decentralized structure of the German energy sector and a shift from conventional to renewable energies. Additionally, current developments in the field of digitalization, e.g., the development of intelligent metering and consumption meters or the connection of decentralized power generators to virtual power plants, new technologies for power storage, or the deep integration of electric vehicles are examples of emerging areas of innovation (Appelrath et al. 2012). With these developments, the energy sector has to manage two transformation processes at once—the energy transition and the digitalization of the energy sector. These two processes not only address parts of this sector but also the whole energy economic value chain and, therefore, cause changes along its value creation process. These reconfigurations from the value chain toward value-added networks trigger adjustments in current BMs of established companies and lead to new possibilities that arise for start-ups through innovative BMs and technologies (Bersch et al. 2014).

This study aimed to analyze the BMs of energy start-ups and to draw a comparison between the developments of BMs in specific market segments and for specific technologies.

4.1 Theoretical Foundation

Recent studies on the analysis of BMs show that, despite different markets, technologies, or ecosystems, there are common market- or technology-related similarities in BMs. As already introduced in the previous chapter, such similarities are used to structure the BMs of companies using patterns. One of the most famous works on these so-called BM Patterns (BMPs) is the studies of the St. Galler BM Navigator by Gassmann et al. (2013a). Overall, they identify 55 different BMPs. According to their studies, over 90% of BM innovations are recombinations of these 55 BMPs. In these studies, a BM is defined by who the customers are, what is sold, how it is managed, and how to generate a profit. For the sake of simplification, Gassmann et al. (2013a) divide the description of a BM into four questions, who-what-how-value.

The questions of "who?" and "what? " address the external dimension of a BM and the questions of "how? " and "value? " the internal dimension. This study used these preliminary works to characterize the BMs of the energy start-ups with the blueprints of the BM Navigator. The simple yet holistic model for the description of the BM by Gassmann et al. (2013a) allowed the identification of the underlying BM using a few selected information.

4.2 Methodology

In close cooperation with Innoloft , a sample of 1.459 international energy start-ups founded between 2000 and 2019 was analyzed (of which 720 were founded in Germany). The sample covers the market segments including Energy generation, Energy supply, Energy storage, Energy market and trade, Energy efficiency and environment, Smart Home, and Electromobility. These segments are again subdivided into several technologies. All start-ups are listed with categorization and description of the technology used as well as information about the BM (target group, value proposition, value creation, and revenue streams). A special feature of the Innoloft start-up Database is the recording of very young "pre-seed" start-ups as well as the available quality of information. The data basis is made possible by Innoloft using a hybrid approach of "network" and "scouting". For Innoloft, "network" means the operation of a digital ecosystem to network between start-ups, investors, and established companies. "Scouting" describes Innoloft's expertise in start-up and technology scouting as well as data science. Innoloft combines the industry and technology knowledge of its employees with modern IT technologies, such as web crawling and neural networks, to intelligently and efficiently collect and classify data on start-ups based on their Internet presence and web activities. The start-up and market data generated in this way were analyzed and allocated according to the BMPs of Gassmann et al. (2013a). This enabled a standardized allocation of the BMs to defined and verifiable patterns.

4.3 Results by Market Segment

The analysis considers two different levels of abstraction. On the one hand, the start-ups are examined at the level of the market segments already mentioned above. On the other hand, concerning the topic of Smart Grids, special attention is paid to the electricity supply and, thus, also directly to the area of Smart Grids.

Concerning the superordinate market segments, 60% of the start-ups use at least one of the five following BMPs: **Direct Selling (overall distribution: 43%):** A company's products are not sold through intermediary channels, but are available directly from the manufacturer or service provider. **Solution Provider (overall distribution: 26%):** A full-service provider offers total coverage of products and services in a par-

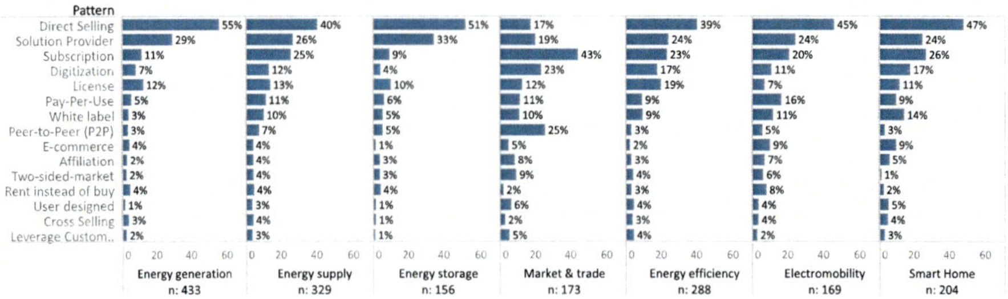

Fig. 8 Breakdown of the 15 most common business model patterns by market segment (multiple answers possible), n = Number of start-ups, Source: own calculation based on the Innoloft database

ticular domain, consolidated via a single point of contact. **Subscription (overall distribution: 21%):** The customer pays a regular fee, typically on a monthly or an annual basis, to gain access to a product or service. **Digitization (overall distribution: 13%):** Offering existing products or services in a digital version, which have advantageous characteristics compared to the physical variant, e.g., in lower production costs, a wider range, or faster distribution. **License (overall distribution: 13%):** Focus on the development of the intellectual property, which is licensed to other companies. The aim is not the own conversion and utilization of knowledge in the form of own products, but rather the commercialization of the usage rights. On average, the start-ups combine 1.9 different patterns, which is reflected in the data in the form of multiple responses. Figure 8 shows the distribution of the BMPs by market segment. The figure only shows the BMPs with an overall usage of more than three percent. Some start-ups can be assigned to several market segments based on their products, services, or overall solutions. This increases the total number of start-ups in Fig. 8 to 1.752.

In the technology and hardware-intensive market segments of Energy generation, Energy supply, Energy storage, and Energy efficiency and environment, the patterns of direct selling and solution providers dominate. In the more service and data-driven segment of the Energy Market and trade, on the other hand, the focus is on subscription, digitization, and peer-to-peer. 43% of start-ups use the subscription pattern as their main revenue stream. Peer-to-peer is used by a quarter of the start-ups and focuses on the gathering of homogeneous groups between which certain services are exchanged. The company itself usually acts as an intermediary or platform. On average, start-ups from this market segment use a combination of 2.3 patterns. The Smart home segment is also characterized by combinations of multiple patterns, with Direct selling also being used by almost half of all start-ups. Due to the combination of software and hardware solutions, which is common in this segment, the Solution Provider and Subscription patterns are also strongly represented here, each accounting for almost 25%. The E-commerce and White label models are also more strongly represented. In the former, a focus is placed on the pure online distribution of solutions. A white label producer allows other companies to sell its products under their brands. The same product is, thus, often sold by several marketers under different names. In the energy-related segment of Electromobility, the Pay-per-use pattern is also more widespread. With this pattern, the actual usage of the service or product is measured and adequately charged. Costumers benefit from the additional flexibility even if priced higher. This pattern is particularly preferred for sharing services and short leases.

4.4 Results from a Electricity Supply Sector Perspective

In the data basis of the Innoloft database, the sub-sector of Electricity supply is subordinated to the market segment of the Energy supply. The database contains entries of 258 start-ups related to Electricity supply and founded between 2000 and 2019. The analysis of this sub-sector considers start-ups that can be assigned to the topics Decentralized energy systems (31%), Industrial power supply (6%), Power electronics (8%), Smart Grid (34%), Power supply for private households (8%), and Supply services (14%). In this area, too, combinations of five different patterns can be used to describe 60% of start-ups. These are the same five patterns for the market segments as a whole, albeit in different distributions: Direct Selling (45%), Subscription (32%), Solution Provider (29%), License (17%), and Digitization (15%). The start-ups here combine 2.3 patterns on average. A more detailed analysis, however, does not show such a clear trend as in the superordinate market segments. Figure 9 shows the 15 most common BMPs for the topics related to Electricity Supply.

The start-ups related to Decentralized power systems usually offer key-ready solutions. This explains the strong focus on the patterns of Direct selling and Solution provider. Just under 60% of start-ups use direct selling and around one in three start-ups acts as a solution provider. These two patterns are often used together. On average, companies in this category use 1.95 different patterns. In contrast, no clear focus can be derived from the Industrial power supply segment. On average, start-ups in this category use 2.5 patterns. One possible explanation for the heterogeneous distribution of patterns is the strong focus on industrial customers and, thus, on individual project business. However, the small study group of only 16 start-ups should also be noted here. In the Power electronics category, direct selling is the dominant method, with 67% of start-ups marketing their solutions directly. In some cases, the solution provider or license model is used. This shows the strong focus of the start-ups on hardware components, which are sold either as components or key-ready solutions directly to the end customer or corresponding integrators. In the case of Power supply for private households, it is striking that start-ups in this category use an average of 4.5 patterns. This results in a rather heterogeneous distribution of patterns without clear dominance. Only the Digitization and E-commerce patterns stand out slightly. According to these patterns, every second start-up in this category uses purely digital solutions and 45% of the start-ups rely on online sales. However, the small group size of 14 start-ups must also be considered here. The Smart Grid category covers a very broad range of different technologies and services such as expansion and planning, operation and monitoring at distribution network and trans-

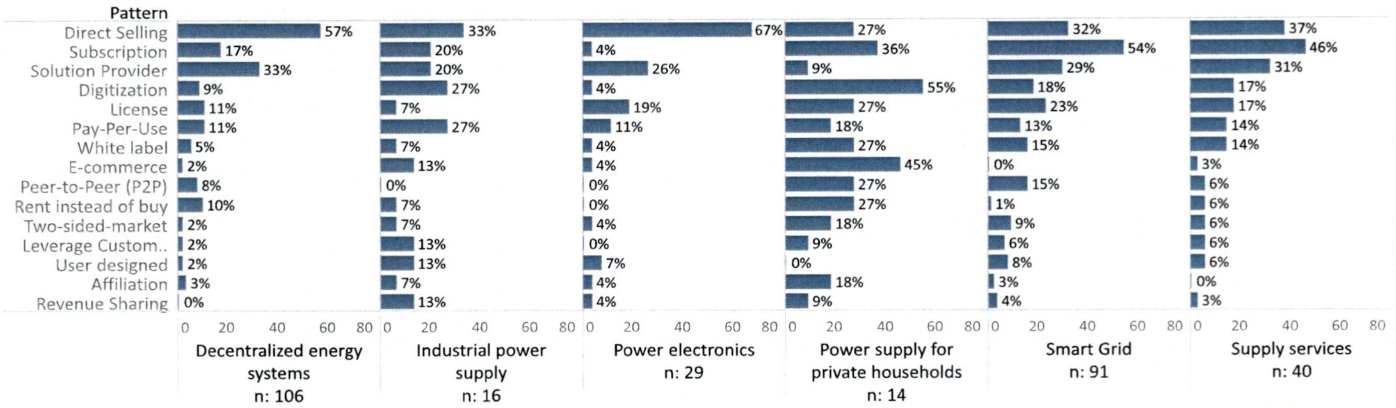

Fig. 9 Breakdown of the 15 most frequent patterns in the electricity supply sector (multiple answers possible), n = Number of Start-ups, Source: own calculation based on the Innoloft database

mission network level, and the provision of corresponding hardware components. On average, the associated start-ups use 2.5 BM patterns for this purpose. Overall, the Subscription pattern is most strongly represented, being used by just over half of all start-ups. This pattern is mainly used by start-ups in the area of monitoring, i.e., where a regular, usually data-driven service is offered. Besides, the combination of Direct selling and Solution provider is again evident here, which is particularly relevant for component manufacturers. The Support Service category covers topics such as metering services, network management, and other system services. Here, too, the Subscription model is in the foreground. But Direct selling, again often in combination with the Solution Provider model, is also widely used. The results of the BM analysis show that many start-ups come into direct contact with customers in the initial phase and, therefore, increasingly use direct sales. This is certainly also related to the strong B2B orientation of many start-ups. Digital tools are increasingly used to increase operational efficiency and reach potential customers. Start-ups with a focus on digital interfaces typically derive part of their value creation from customer data and generate a large part of their revenues from subscriptions. Many start-ups in the various segments are also increasingly focusing on the provision of a key-ready solution, often a combination of software and hardware, to be available to customers as a solution provider as a central point of contact. Understanding the BMs of energy start-ups is an important element to gain deeper knowledge about the ongoing transformation processes in the energy industry. This also includes the extent to which trends such as digitization, user-centered services, home automation, or peer-to-peer companies have already penetrated the energy sector. This will make it possible to forecast developments more accurately and to derive better recommendations for action for setting incentives and overcoming obstacles in the area of start-ups for politicians.

Review Questions

- Define the term Business Model in your terms.
- Describe the goals and areas of application for BMs.
- Describe the steps of the BM Design Process.
- Fill out the BM Canvas for one of the case studies in the next chapter.
- Describe one of the case studies using the introduced business model patterns.
- Control questions: Why do you observe such a start-up dynamic in the energy sector? Explain the BM dimensions. Explain the phases of business model design.

References

Al-Debei MM, Avison D (2010) Developing a unified framework of the business model concept. Eur J Inf Syst 19(3):359–376

Al-Debei MM, El-Haddadeh R, Avison D (2008) Defining the business model in the new world of digital business. AMCIS 2008 proceedings

Alt R, Zimmermann H-D (2001) Introduction to special section - business models. Electron Mark Int J 11(1):1019–6781

Amit R, Zott C (2001) Value creation in e-business. Strateg Manag J 22(6–7):493–520

Andersson B, Bergholtz M, Edirisuriya A, Ilayperuma T, Johannesson P, Gordijn J, Grégoire B, Schmitt M, Dubois E, Abels S, Hahn A, Wangler B, Weigand H (2006) Towards a reference ontology for business models. Conceptual modeling - ER 2006. Springer, Berlin Heidelberg, pp 482–496

Appelrath H-J, Kagermann H, Mayer C (2012) Future energy grid. Springer, Berlin Heidelberg, Berlin, Heidelberg

Bach N, Buchholz W, Eichler B (2003) Geschäftsmodelle für wertschöpfungsnetzwerke — begriffliche und konzeptionelle grundlagen. In: Geschäftsmodelle für Wertschöpfungsnetzwerke. Gabler, pp 1–20

Baden-Fuller C, Morgan MS (2010) Business models as models. Long Range Plan 43(2–3):156–171

Ballon P (2007) Business modelling revisited: the configuration of control and value. Info 9(5):6–19

Barney JB (2010) Gaining and sustaining competitive advantage, 4 edn. Pearson

Barney JB, Arikan AM (2001) The resource-based view: origins and implications. The blackwell handbook of strategic management. Oxford, pp 124–188

Bellman R, Clark CE, Malcolm DG, Craft CJ, Ricciardi FM (1957) On the construction of a multistage, multi-person business game. Oper Res 5(4):469–503

Bersch J, Gottschalk S, Muller B, Niefert M (2014) The mannheim enterprise panel (mup) and firm statistics for germany. Zentrum für europäische Wirtschaftsordnung. Discussion Paper No 14–104

Bieger T, Reinhold S (2011) Innovative Geschäftsmodelle, Chapter Das wertbasierte Geschäftsmodell – Ein aktualisierter Strukturierungsansatz. Springer, Berlin, Heidelberg, pp 13–70

Bouwman H (2002) The sense and nonsense of business models. Int Workshop Bus Models

Burkhart T, Krumeich J, Werth D, Loos P (2011) Analyzing the business model concept - a comprehensive classification of literature

Casadesus-Masanell R, Ricart JE (2011) Geschäftsmodelle richtig gestalten. Harv Bus Manag 33(5):52–61

Chesbrough H (2002) The role of the business model in capturing value from innovation: evidence from xerox corporation's technology spin-off companies. Ind Corp Chang 11(3):529–555

Clauss T (2017) Measuring business model innovation: conceptualization, scale development, and proof of performance. R&D Manag 47(3):385–403

Deelmann T, Loos P (2003) Visuelle methoden zur darstellung von geschäftsmodellen - methodenvergleich, anforderungsdefinition und exemplarischer visualisierungsvorschlag. Work Pap Res Group Inf Syst Manag

Deelmann T, Loos P (2004). Vorschlag zur grafischen reprasentation von geschaftsmodellen. Work Pap Res Group Inf Syst Manag

Demil B, Lecocq X (2010) Business model evolution: in search of dynamic consistency. Long Range Plan 43(2–3):227–246

Doganova L, Eyquem-Renault M (2009) What do business models do? Res Policy 38(10):1559–1570

Doleski OD (2015) Integrated Business Model - Applying the St. Gallen Management Concept to Business Models. Gabler

Engelken M, Römer B, Drescher M, Welpe IM, Picot A (2016) Comparing drivers, barriers, and opportunities of business models for renewable energies: a review. Renew Sustain Energy Rev 60:795–809

Eriksson H-E, Penker M (2000) Business patterns with UML. Wiley
Eurich M, Weiblen T, Breitenmoser P, Boutellier R (2013) A 'networked thinking' approach to business model design. In: The proceedings of The XXIV ISPIM conference - innovating in global markets: challenges for sustainable growth, pp 1–13
Evans JD, Johnson RO (2013) Tools for managing early-stage business model innovation. Res-Technol Manag 56(5):52–56
Foss NJ, Saebi T (2018) Business models and business model innovation: between wicked and paradigmatic problems. Long Range Plan 51(1):9–21
Frankenberger K, Weiblen T, Csik M, Gassmann O (2013) The 4i-framework of business model innovation: a structured view on process phases and challenges. Int J Prod Dev 18(3/4):249
Freeman RE (2010) Strategic management: a stakeholder approach. Cambridge university press
Gassmann O, Frankenberger K, Csik M (eds) (2013) Geschäftsmodelle entwickeln. Carl Hanser Verlag GmbH & Co. KG, München
Gassmann O, Frankenberger K, Csik M (2013) Geschäftsmodelle entwickeln. In: Gassmann O, Frankenberger K, Csik M (eds), Geschäftsmodelle entwickeln. München: Carl Hanser Verlag GmbH & Co. KG , pp I–XIV
Gassmann O, Frankenberger K, Csik M (2017) Geschäftsmodelle entwickeln. Carl Hanser
Ghaziani A, Ventresca MJ (2005) Keywords and cultural change: frame analysis of business model public talk, 1975–2000. Sociol Forum 20(4):523–559
Hamel G (2002) Leading the revolution: how to thrive in turbulent times by making innovation a way of life. PLUME, New York
Hax AC, Majluf NS (1991) Strategisches management. Ein integratives Konzept aus dem MIT, Campus
Heinemann H (2008) Erfolgsfaktoren bei Unternehmensgründungen. GRIN
Heuser L, Lacher S, Perlmann S (2007) Flexible prozessgestaltung als basis innovativer geschaftsmodelle - von der service-orientierten architektur zur vision des business webs. Wirtschaftinformatik Proc 2007:4
Jolink A, Niesten E (2013). Sustainable development and business models of entrepreneurs in the organic food industry. Bus Strategy Environ 24(6):386–401
Jonda M (2004) Szenario-Management digitaler Geschaftsmodelle - Skizze einer Geschaftsmodellierung am Beispiel von Mobile-Health-Dienstleistungen. PhD thesis, Carl von Ossietzky Universitat Oldenburg
Jones GM (1960) Educators, electrons, and business models: a problem in synthesis. Account Rev 35(4):619
Joyce A, Paquin RL (2016) The triple layered business model canvas: a tool to design more sustainable business models. J Clean Prod 135:1474–1486
Kagermann H, Österle H (2007) Geschäftsmodelle 2010: Wie CEOs Unternehmen transformieren. Frankfurter Allgemeine Buch
Kallio J, Tinnilä M, Tseng A (2006) An international comparison of operator-driven business models. Bus Process Manag J 12(3):281–298
Karakaya E, Nuur C, Hidalgo A (2016) Business model challenge: lessons from a local solar company. Renew Energy 85:1026–1035
Karami M, Madlener R (2018) Business model innovation for the energy market: joint value creation for electricity retailers and their residential customers. SSRN Electron J
Kittner N, Lill F, Kammen DM (2017) Energy storage deployment and innovation for the clean energy transition. Nat Energy 2(17125)
Kley F, Lerch C, Dallinger D (2011) New business models for electric cars—a holistic approach. Energy Policy 39(6):3392–3403
Kollmann T, Hensellek S (2016) The e-business model generator. In: Encyclopedia of e-commerce development, implementation, and management. IGI Global, pp 26–36
Krumeich J, Burkhart T, Werth D, Loos P (2012) Towards a component-based description of business models: a state-of-the-art analysis. In: AMCIS 2012 proceedings

Lambert S (2003) Making sense of business models. In: School of commerce research paper series: 03-10
Lambert S (2008) A conceptual framework for business model research. BLED 2008 Proc, 24
Laurischkat K, Viertelhausen A (2017) Business model gaming: a game-based methodology for e-mobility business model innovation. Procedia CIRP 64:115–120
Lindgren P, Rasmussen OH (2013) The business model cube. J Multi Bus Model Innov Technol 1(3):135–180
Lindström C-G (1999) Lessons learned from applying business modelling: exploring opportunities and avoiding pitfalls. Perspect Bus Model. Springer, Berlin Heidelberg, pp 151–164
MacInnes I, Hwang J (2003) Business models for peer to peer initiatives. In: 16th bled eCommerce conference - eTransformation, pp 10
Madina C, Zamora I, Zabala E (2016) Methodology for assessing electric vehicle charging infrastructure business models. Energy Policy 89(C):284–293
Magretta J (2002) Why business models matter. Harvard Bus Rev 80:86–92, 133
Malek M, Ibach PK, Ahlers J (2004) Entrepreneurship par excellence. Dpunkt
Maurya A (2012) Running lean - iterate from plan A to a plan that works (lean series). O'Reilly
McGuire JW (1965) How much freedom does business really want? Bus Horiz 8(2):73–78
Meinhardt Y (2002) Veränderung von Geschäftsmodellen in dynamischen Industrien. Gabler
Morris M, Schindehutte M, Allen J (2005) The entrepreneurs business model: toward a unified perspective. J Bus Res 58(6):726–735
Nagl A (2010) Der Businessplan. Springer
Onetti A, Zucchella A, Jones MV, McDougall-Covin PP (2012) Internationalization, innovation and entrepreneurship: business models for new technology-based firms. J Manag Gov 16(3):337–368
Osterwalder A, Pigneur Y (2002) An ebusiness model ontology for modeling ebusiness. In: 5th bled electronic commerce conference ereality: constructing the eeconomy, pp 2
Osterwalder A, Pigneur Y (2010) Business model generation. A handbook for visionaries, game changers, and challengers. Wiley
Osterwalder A, Pigneur Y, Tucci CL (2005) Clarifying business models: origins, present, and future of the concept. Commun Assoc Inf Syst 16
Pateli AG, Giaglis GM (2004) A research framework for analysing eBusiness models. Eur J Inf Syst 13(4):302–314
Porter ME (2004) Competitive strategy: techniques for analyzing industries and competitors. Free Press, New York
Rentmeister J, Klein S (2003) Geschäftsmodelle — ein modebegriff auf der waagschale. In: Die Zukunft des Electronic Business. Gabler, pp 17–30
Reuver Md, Bouwman H, MacInnes I (2009) Business models dynamics for start-ups and innovating e-businesses. Int J Electron Bus 7(3):269
Richardson J (2008) The business model: an integrative framework for strategy execution. Strateg Chang 17(5–6):133–144
Roome N, Louche C (2015) Journeying toward business models for sustainability. Organ Environ 29(1):11–35
Schallmo DRA, Brecht L (2010) Business model innovation in business-to-business markets - procedure and examples. Proc 3rd ISPIM Innov Symp Manag Art Innov Turn Concepts Real 12:15
Scheer C, Deelmann T, Loos P (2003) Geschaftsmodelle und internetbasierte geschaftsmodelle - griffsbestimmung und teilnehmermodell. In: Working papers of the research group information systems & management
Schoegel K (2001) Geschäftsmodelle?: Konstrukt-Bezugsrahmen-Management. FGM, München
Seddon PB, Lewis GP (2003) Strategy and business models: what'the difference? In: 7th pacific asia conference on information systems, pp 17
Shafer SM, Smith HJ, Linder JC (2005) The power of business models. Bus Horiz 48(3):199–207
Singler A (2012) Businessplan, 3rd edn. Haufe-Lexware

Stähler P (2002) Geschäftsmodelle in der digitalen Ökonomie. Merkmale Strategien und Auswirkungen. Josef Eul
Teece DJ (2010) Business models, business strategy and innovation. Long Range Plan 43(2–3):172–194
Thommen J-P (2007) Betriebswirtschaftslehre, 7th edn. Versus
Wirtz B, Göttel V, Daiser P (2016) Business model innovation: development, concept and future research directions. J Bus Models 4(1) (2016): 2016 open issue–
Wirtz BW (2010) Business model management. Design - Instrumente - Erfolgsfaktoren von Geschäftsmodellen. Gabler
Wirtz BW, Kleineicken A (2000) Geschäftsmodelltypologien im internet. WiSt - Wirtschaftswissenschaftliches Studium 29(11):626–634
Zott C, Amit R (2007) Business model design and the performance of entrepreneurial firms. Organ Sci 18(2):181–199
Zott C, Amit R (2008) The fit between product market strategy and business model: implications for firm performance. Strateg Manag J 29(1):1–26
Zott C, Amit R, Massa L (2011) The business model: recent developments and future research. J Manag 37(4):1019–1042

Case Studies in the Smart Grid Sector

Clemens van Dinther, Markus Lau, and Orestis Terzidis

Learning Objectives

- Understand and analyze the case study,
- Understand the market situation and classify the resulting opportunities and risks,
- Apply the theoretical approaches and model from previous chapters on real-world cases,
- Identify value propositions,
- Describe a value architecture,
- Understand a revenue stream, and
- Answer business-related questions.

1 Introducing Case Studies

The changes at the technical, regulatory, and market levels discussed in the book enable the emergence of new business models. In particular, new opportunities arise from the use of advances in digitization for applications in the smart grid. In this

C. van Dinther (✉)
ESB Business School, Reutlingen University, Alteburgstr. 150, 72762 Reutlingen, Germany
e-mail: clemens.van_dinther@reutlingen-university.de

M. Lau · O. Terzidis
Institute for Entrepreneurship, Technology Management and Innovation (EnTechnon), Karlsruhe Institute of Technology (KIT), Fritz-Erler-Straße 1-3, 76133 Karlsruhe, Germany
e-mail: markus.lau@kit.edu

O. Terzidis
e-mail: orestis.terzidis@kit.edu

C. Dinther et al. (eds.), *Smart Grid Economics and Management*,
Lecture Notes in Energy 51, https://doi.org/10.1007/978-3-030-84286-4_9

chapter, we would like to present and analyze examples of such new business models in two case studies. In recent years, a number of new companies have been established in the market. For example, in the B2B business, the German start-up Entelios and the American start-up EnerNOC successfully developed solutions in the area of demand response/demand-side management and were later acquired by the Italian energy company ENEL after a merger. With a growing share of renewable generation, especially forecasting of generation output of wind and PV plants is gaining importance. The German company Energy & Meteo Systems has specialized in this service. With Younicos, a German-American start-up has developed control software for electricity storage, targeting flexible management of electricity supply and demand in the distribution grid. Younicos was acquired in 2017 by Aggreko, a Scottish energy company. Discovergy is a company in the B2C market that offers smart metering services for end customers. These examples show just a sample of new areas where demand for new solutions has emerged. It also shows that new business developed in both sectors, B2B and B2C, and that successful businesses are sometimes overtaken by established players in the market. The transition of the energy system is progressing, and we will certainly see more start-ups and business models emerging. We cannot provide a complete overview within the scope of this book. Instead, we present Power Plus Communications (PPC) and GridX, two companies that can be used to elaborate on the challenges and opportunities of new business models.

2 Case Study on Power Plus Communications (PPC)

You are a market observer and tasked to evaluate the German company Power Plus Communications (PPC)[1] and its business prospects. PPC, which provides data communication systems for smart metering and smart grids, is a specialist for ICT-Solutions and market leader for certified smart meter gateways according to the German protection profile for privacy and security. Smart Meter Gateways are the central communication entity that is responsible for both, meter data communication as well as access to controllable devices of the household.

Today, with the increasing use of renewable energy sources and a greater need for energy grid flexibility, utility providers are looking to implement smart grid technologies and use data to improve the performance of the distribution grid. However, building a robust smart grid presents utility providers with a variety of challenges and creates a need for external solutions. One major requirement of a smart grid is collecting the necessary data in the field (e.g. meter data and sensors at substations) and transmitting it to the utility providers so that they can monitor the actual grid status, analyze it, forecast the grid capacity utilization, and take necessary actions to control the grid (e.g. switch on/off devices or generation). This is in direct contrast to how utility providers have historically operated: Previously, utility providers did

[1] Dudenstraße 6, 68167 Mannheim, Germany, https://www.ppc-ag.de.

not need, and therefore, did not install sensors or meters for real-time data collection. With the smart grid utility, providers can set up an infrastructure to measure on a real-time basis (at consumption or feed-in points) and to establish a bi-directional information flow to manage power supply and protect the grid from outages, overloads, overvoltages, etc. As a result, utility operators must implement intelligent metering systems and a robust communication infrastructure. Certainly, the communication infrastructure must be protected against cyber-attacks, since the electricity grid is a critical infrastructure. Building a communication infrastructure that supports utility operators' smart grid needs can be prohibitively expensive and difficult, however. This creates new responsibilities to comply with strict data privacy and security requirements laid down with the European General Directive on Data Protection (GDPR) and Germany's Federal Data Protection Law (BDSG) as well as with Metering Point Operating Law (MsbG) among other pieces of legislation that affect this highly regulated industry. Building a robust, cyber-secure smart grid, compliant with all these regulations, presents utility providers with a variety of challenges, and creates opportunities for third-party companies with the technical expertise utilities don't traditionally have. Technologies to implement bi-directional communication in a smart grid are broadband powerline (BPL) and mobile communication technology. BPL uses the already existing low- and medium-voltage power grid to transmit data. Because it utilizes already-existing infrastructure, this solution has several benefits:

1. This communication technology can be used to transmit data from any consumption point within an energy grid. Because all consumption points are by definition connected to the energy grid, they can all be connected with this technology. This is true even in less developed countries that may lack key infrastructure.
2. Using existing infrastructure cuts down on implementation and maintenance costs for utility providers.

Utility providers usually do not have much experience in turning their energy grids into a broadband communication network though and therefore seek out external suppliers and partners. Mobile technology is similarly already well developed and doesn't require utility providers to build an entirely new communication network, thus saving on investment costs (with higher operating cost due to subscription fees) and increasing flexibility. The European Union requires the member states to invest in intelligent metering infrastructure and to equip 80% of meter points with intelligent metering systems (IMS). As the EU does not set a standard for the IMS, it is up to the member states' regulation to set the standard. Germany has more than 40 million meter points in total as from about 2020 on regulation requires every customer with an energy consumption of more than 6,000 kWh to be equipped with an IMS. By 2032, at least 80% of the meters should be an IMS. In order to secure the energy system as a critical infrastructure against cyber-attacks and to meet data protection requirements, the regulation has laid down high-security requirements for an IMS, which includes two components: a smart meter and a smart meter gateway (SMG), a device responsible for securing meter data communication and access to controllable devices in the household. Consequently, every device in the IMS has to be certified by the Federal Office for Information Security according to a protection profile. The process of certification is tough and takes a number of years: By end of 2018, PPC was the only company with a smart meter gateway ready for certification by the agency. But competitive regulations required that three SMGWs had to be available on the market. By the end of 2019, two further companies were able to successfully conclude the certification process. Besides meeting regulatory requirements, smart meter gateways should also be compatible with the hardware (meters, communication technologies) and software (administration systems) of many other companies in the market. Although all market requirements were met as requested by the regulator, IMS rollout starts on a moderately low level in 2020. As a German company, PPC not only primarily targets European DNOs and utilities but also has customers outside Europe. Its customers include all leading German utilities and network operators. Consequently, the company has a strong brand as a market leader in the segment. Located in Mannheim, PPC benefits from a central location with a direct connection to many European cities. PPC shareholders include the management, members of the supervisory board, and employees of PPC. In addition, a couple of family offices and the venture capital fund Climate Change Capital have shares in PPC. PPC offers two main categories of products and services to its customers. On the product side, PPC sells technologies that support data communication for smart metering and smart grids, including smart meter gateways. On the service side, PPC offers consulting and project management as well as training and workshops. PPC's products aim to offer utility providers cost-efficient, scalable, and flexible ways to implement their smart grids. The services that PPC provides are aimed at guiding customers through the complexities of smart grid communication and helping them successfully implement and use PPC's products.

PPC relies on its employees' knowledge and expertise in the BPL and mobile communication realm to create value for clients. The company is compact, with 80 employees working as project managers, network designers, technical developers, solution managers, software developers, and IT employees. These employees bring knowledge and expertise about BPL and mobile communication technology that utility providers lack but require for the implementation of their smart metering and smart grid projects. The company is *fablos*, that is, it does not have its own manufacturing facilities; its products are built by third-party manufacturers. The two founders, Ingo Schönberg and Eugen Mayer, PPC's two managing directors, have both led the company since its inception in 2001. PPC's ecosystem of partners includes not only resellers and suppliers but also research and development organizations, technical standards organizations, and regulatory authorities. The company is an active participant in many research and development projects, where they cooperate with research institutes and other organizations to develop new solutions to industry challenges. Outside of Germany, PPC works with resellers including Swistec (Switzerland), Mikronika (Poland), and CleverPower (Czech Republic).

Review Question

- Who are PPC's clients?
- What challenges do the clients face?
- How does PPC help in solving these challenges?
- Which other stakeholders profit from PPC's solutions?
- Is PPC's business model new in the energy sector?
- What role does the decarbonization of the energy sector play with respect to PPC's further prospects?
- Why is PPC's business model new in the energy sector? Which role does the decarbonization of the energy sector play with respect to PPC's further prospect?
- Which value proposition does PPC offer?
- What is PPC's value architecture with respect to resources, competences, internal/external activities, and processes?
- Of which components does the revenue stream consist?
- Has PPC established a value network?
- How do you overall evaluate PPC's business prospect?

3 Case Study on GridX

You're a member of the supervisory board of the German start-up company gridX[2] founded in 2016 by two alumni of RWTH Aachen. Today, you meet with the founders to discuss whether gridX should change its business model (BM). As an experienced tech professional, you are asked to give them advice on if and how they should act.

gridX developed a hardware device, the gridBox, and a dashboard, the gridX app. The gridBox enables the connection of all DERs, independent of the manufacturer. This includes not only various smart home devices but also devices such as heat pumps and battery storage. The main functionality of the gridBox is to connect these multiple hardware devices from different manufacturers to one device and manage all of them with one holistic service. This is necessary because different manufacturers, especially from China and the US, use different protocols that cannot communicate with each other. The gridBox functions as a gateway between the different hardware devices owned by the end-user and the gridX app. The dashboard analyzes and processes the internet of things (IoT) data provided by the gridBox, which can be used by the user to manage their devices. The app also proposes ways to minimize energy usage and can therefore be of economic value. Lastly, it can also warn the user in case of faulty equipment even before it stops working completely. The box is sold directly to end-users for a retail price of 499, which includes the physical hardware

[2] Oppenhoffallee 143, 52066 Aachen, Germany, https://de.gridx.ai/.

and the right to use the app. This price needs to cover the cost of manufacturing the box externally and also the development and maintenance of the app.

The customers (end-users) buy the product to have an easy and intuitive way to connect their smart home devices, yet some of them thought the price was too high. In two years, gridX was able to sell more than 1000 gridBoxes to end-users. The start-up has grown fast and has more than 25 employees now. The team around gridX consists of talents and experienced specialists from leading companies such as Google, McKinsey, and universities such as MIT, Harvard, and RWTH Aachen University. The company relies on its employees to engineer, program, and design solutions across the full stack of technologies. They were also able to secure a Series A investment, having collected several million Euros from well-known investors such as Innogy Ventures, Coparion, and VitoONE. They now propose a change from B2C to B2B business. The changing nature of the energy industry has created an imperative for utility providers to digitally transform themselves. The IoT has enabled the collection and utilization of data at all points across the energy value chain. With 1.7 million solar-energy systems and 70,000-plus EVs in Germany, the opportunity is ripe for new business models and product and management options. In order to fully take advantage of these IoT opportunities, utility providers need to implement IoT cloud infrastructure that enables digital business models along the entire value chain. For this reason, multiple utility providers have contacted gridX about their gridBox. Their first B2B customer is Viessmann, a leading international manufacturer of energy (heating) systems.

gridX has developed a platform which is built on a stack of technologies. At the uppermost level, gridX offers prefabricated solutions for the energy IoT sector. Companies who purchase gridX's platform service have access to a management dashboard, cloud hosting, and technical support for their customers. With this package of services, utility providers are able to easily and rapidly implement IoT solutions without the need to invest massive amounts of time or money into developing their own solutions or managing the technical side of their digital business models themselves. Going one level deeper in gridX's stack of technologies, these services are based on data stored gridX-Cloud, which receives and processes the large amounts of data from the fleet of remote IoT devices that the gridBox connects to. gridX's device management technology enables users to monitor the health of their devices and control their devices remotely, pushes alerts if something goes wrong, and pushes new software updates over the air. At its core, gridX's easily scalable platform is enabled by gridX's physical product: the gridBox communication gateway and its gridOS operating system.

The following Use Cases are possible: gridX Independent Homes solution is most closely tied with the original B2C gridBox offering: It's a white-label product for utility providers that enables flexible energy management and energy monitoring of private households. Households with solar panels and storage batteries can drastically reduce their energy costs and CO_2 footprint as gridX takes weather conditions into account and utilizes the storage capacity and the energy generation peaks. For utility providers, this solution supports grid resiliency and stability, opens the door to future revenue models, and enables up- and cross-selling business models based on data collected by IoT devices. gridX's second solution enables smart charging for EVs. With this solution, which is available for both home and public charging, electric vehicle (EV) charging is made smart by considering the actual market price of energy and the source of energy generation. This information provides power distributors with more detailed information about significant power consumption patterns of their customers, which is useful for planning the required amount of energy correctly and thus minimizing the difference between demand and supply. Smart EV charging can also be connected with the Independent Homes solution so that households can further maximize self-consumption and save costs.

gridX's Microgrids solution allows for temporary independencies of the grid using decentralized energy units. It enables the monitoring, managing, and billing of customer energy flows to prepare forecasts for the management of the decentralized power plant fleets. It also enables peer-to-peer energy trading between consumers. The fourth gridX solution is called Smart Commercial, a holistic energy monitoring and management specifically targeting all companies in industry and commerce that generate their own electricity via a photovoltaic system or a combined heat and power (CHP). Among all four solutions, gridX is offering predictive maintenance. IoT measures energy consumption, temperature, noise, and other indicators to predict malfunctioning systems or machinery. It therefore enables agility and process flexibility, while keeping downtime costs to a minimum. All current gridX offerings are based on their Energy Management System (EMS) which offers many more possibilities and solutions customizable depending on customer needs and requirements. Focusing on B2B enables gridX to serve end-users it otherwise could not, take advantage of other energy firms' direct relationships with end-users, and create value for energy producers as well as energy consumers. The pricing model also differs for the two offerings: Although the gridBox was available to customers with a one-time purchase, the B2B platform service model is subscription-based. This B2B model also expands what gridX can sell to clients and how much revenue gridX can make.

Review Question

- What challenges can be solved by gridX products and services?
- What is gridX's value proposition?
- How does gridX transform their business model by offering B2B services? What effect does this transformation have on the value architecture?
- What do you advice gridX from a technical point of view? How could they benefit from rearranging their business model? Which role does interconnectivity play in this context?
- Who else could profit from gridX's value proposition?
- Who is part of gridX's value network and what kind of partners do you recommend to add?